CHARACTERISTIC TOWNS
特色小镇

思纳史密斯集团 编著

大连理工大学出版社

图书在版编目(CIP)数据

特色小镇 / 思纳史密斯集团编著. — 大连：大连理工大学出版社，2018.2
ISBN 978-7-5685-1149-0

Ⅰ.①特… Ⅱ.①思… Ⅲ.①小城镇—城市规划—建筑设计 Ⅳ.①TU984

中国版本图书馆CIP数据核字（2017）第296255号

出版发行：大连理工大学出版社
　　　　　（地址：大连市软件园路80号　邮编：116023）
印　　刷：上海锦良印刷厂
幅面尺寸：235mm×310mm
印　　张：16.5
字　　数：260千字
插　　页：4
出版时间：2018年2月第1版
印刷时间：2018年2月第1次印刷
责任编辑：张　泓　初　蕾　裘美倩
责任校对：王秀媛　韩然松
封面设计：君誉文化
策　　划：君誉文化

ISBN 978-7-5685-1149-0
定　　价：298.00元

电　　话：0411-84708842
传　　真：0411-84701466
邮　　购：0411-84708943
E-mail：jzkf@dutp.cn
URL：http://dutp.dlut.edu.cn

本书如有印装质量问题，请与我社发行部联系更换。

我们策划、规划的新城和小镇在不断增加，

得益于铺天盖地的发展计划，

不能判断这是好是坏，

只能说这是一个时代的标记，

我们既然在其中幸运地成为它们的设计者，

那就好好去设计吧。

朱轶俊

2017.3.20

Here on the increase are the new cities and towns designed,

Thanks to the development programs in multitude,

It is still too early to judge whether it's bad or good,

Or only a sign of the times,

Lucky as we are to be their designers,

Then go to do the best we could.

Zhu Yijun

2017.3.20

思纳设计

　　思纳设计是思纳史密斯集团在中国及周边国家和地区的项目组织者和设计小组，是集团中国设计和管理体系的最高管理者和决策者，是目前中国规模较大的综合类设计公司之一。

　　思纳设计是规划建筑和旅游行业的产品策划与设计专家，也是在上海为数不多的建筑和规划国家双甲资质单位。思纳设计在智慧模块建筑、城市和旅游产业策划、室内装饰和艺术设计、环境景观和能源分析、特色酒店跨界组合运营规划等领域都有卓越的人才及成功案例。我们强调用技术和艺术的手段支持不同的客户对市场的需求，并善于通过策划性的规划和设计将城市乡村、居住旅游、商业娱乐和主题产业融合发展，开拓性地创造了帐篷客生态主题酒店、智慧单元建筑、景区运营和产业协同发展的乡级产业小镇等规划运营的产品专利模板，并不断地结合实际运营的绩效进行修正和发展。

　　现在思纳设计已成为可以在中国提供建设项目全过程服务的综合设计咨询和管理集团，我们相信客户的事业便是我们共同的事业，我们将全力以赴，努力工作，实现梦想。

CNAS DESIGN

CNAS Design is the project organizer and the design team of CNASMITH group in China and the surrounding countries and areas. Being the group's top management and decision maker in design and management system in China, it is currently one of the large comprehensive design company in China.

CNAS Design is one of the product planning and design professionals in the construction and tourism industries, as it has been acknowledged in Shanghai as one of a handful of the state-double-A-qualification in architecture and planning. CNAS Design has a rich pool of talents and successful cases in a wide range of fields such as smart module architecture, city planning and tourism planning, interior decoration and art design, landscape design and energy analysis, cross-border operation of characteristic hotel, etc. We spare no effort to support different customers' market needs by applying technological and artistic approaches, and we are particularly proficient at integrating the city and the village, residence and tourism, commercial entertainment and theme industry by means of strategically planning and designing. Furthermore, we have pioneered to create patent product templates for planning and operation, which are constantly modified and developed in accordance with the actual operational efficacy. The products that can apply our patented templates include tent off ecological theme hotel, intelligence unit construction, industrial town with coordinated development of scenic spot and industry.

Presently, CNAS Design has emerged into a comprehensive design consulting and management group that can provide the entire process of construction projects in China. We believe that the customers' cause is also our cause, and we are bent on making every possible effort to realize our common goal.

序言 PREFACE

特色小镇，根在产业
——思纳设计董事长朱轶俊谈小城镇规划

文：赵夏榕
翻译：云舒

一条萦绕着丁香气味的雨巷，一组绵延相依的青石台阶，一间古意盎然的夯土老屋里探出的几支早梅……

说起小城镇，人们心里总是有着许多美好的画面。它既是乡愁、是历史，也属于生活、属于当代。随着城市建设步伐日趋稳健，小城镇不仅吸引着人们的心灵回归，更将成为未来经济发展的极重要一环。

作为涉足特色小城镇规划设计的先行者，思纳设计在过去的八年中深耕特色小镇、美丽乡村规划设计领域。近年来，更是自己投资进行小镇项目的开发与运营。思纳设计的董事长朱轶俊先生，多年来遍访中国的美丽小镇，积累了丰富的一线项目设计经验。

他认为"特色小镇"的"特色"不仅在于它的布局、空间形态，还在于它的产业特色。只有产业策划与传统意义上的区域规划设计相配合，才能成为指导性的方案。在特色产业的导入方面，思纳设计也有深入的研究——除了较大的产业之外，更注重对于当地"微产业"的推动，通过成立"帮帮基金"令当地微小商业焕发新貌。

于是，那些保留着历史记忆的小镇便得以更美丽的形态，带着活力迎接未来。

产业新城、美丽乡村、特色小镇正待发力

当您走进一个小镇，会被其中的什么东西打动？这些体验和感受，对您的设计有什么样的影响？

朱轶俊：在过去的20年里，我走访了二三百个不同类型的小镇，其中很多都让我印象深刻。比如，当年我看完电影《廊桥遗梦》之后，就去寻找中国的"廊桥"结果发现西塘就很美；再比如日本的温泉小镇虹夕诺雅(HOSHINOYA)，其实我们国内也有类似的地方。

小镇，总是寄托着人的某种美好的想象。对我来说，我可能还会把记忆中美好的东西以及游历中的美好片段拼到一起——目前思纳设计在湖州南浔投资的一个小镇项目就是这样做的。

FEATURED TOWN, ROOTED IN INDUSTRY
——Town planning interview with Zhu Yijun, the Chairman of CNASMITH

Text: Zhao Xiarong
Translation: Lilac

A rain lane carrying the aroma of blooming lilacs, a set of continuous bluestone steps, several plum blossoms lean out the quaint rammed earth house…

When talking about towns, many beautiful images will come into people's mind. Town means not only nostalgia and history, but also life and present. After the urbanization process, towns are becoming the house sheltering people's heart and the significant part of economic development in the future.

As the pioneer of featured town planning, CNASMITH has been involved in the planning of featured town and country for eight years with projects all over China. Recently, it also has invested in the developing and operation of towns. Zhu Yijun, the Chairman of CNASMITH has visited towns all over China in the past years and accumulated rich experience in projects planning.

In his opinion, the characteristics of featured towns includes not only its layout and spatial form, but also its industry features. Good planning only comes out from industry planning together with space planning. CNASMITH has made intensive study of specialty industries. It enhances the development of "micro-industry" on the base of larger industries and helps mini-size industries to have a new look by "Bangbang Fund".

As a result, the towns with the historic memories are embracing the future in a better way.

New industrial town, beautiful country and featured towns are about to blossom

What impressed you most when you are walking into a town? And what are the influences caused by these experiences?

Zhu Yijun: I have visited about 200 to 300 towns of different types in the past 20 years, which are mostly impressive. For example, after I saw the film *The Bridges of Madison County*, I started out to find more in China and then found the beauty of Xitang. And similar towns can be found in China like the hot spring town Hoshinoya in Japan.

Towns are always with the best imaginations. And for me, I may blend the best points in my memory with the scenes I have seen in towns. That's what we have done for the town project in Nanxun, Hunan we have invested recently.

您是从什么时候开始从事小城镇规划与设计的？当时您对于相关领域的发展趋势、规划设计的理念有怎样的判断？为什么？

朱轶俊：我们从八年前开始做小城镇和美丽乡村规划，其实这里面包括了产城融合、美丽乡村、特色小镇系列项目。

做这些项目，是因为我们发现中国城市的建设基本上已经到了饱和状态——毕竟中国大部分的土地是在农村和城镇，它们的发展前景广阔。接下来的五年、十年中，围绕着城市周边的小城镇改造和升级、卫星城市的建设，小到美丽乡村，大到跟产业相关的一些特色小镇，都是中国城乡建设的重要内容。

回顾这段历程，您认为规划设计的背景、任务等有什么样的改变？

朱轶俊：从八年前到现在，我们经历了从摸索到实践、提升，再到发展这样几个递进式的阶段。

八年前开始做这些事情的时候，大家并不清楚到底要做什么，只是把在城市里面做的一些东西（住宅、产业园）放到离城市比较近且交通方便的地方去，并导入一些新的业态。后来我们发现，当城市发展到了一定的程度以后，出现了"城市病"，再加上空气污染给人们心态带来的变化，让人们对郊外和生态环境良好的区域产生向往，这种向往继而又转化成实际的需求。同时，随着互联网和物联网的发达，使异地办公、异地生产有了可能。于是，首先出现了将城市周边交通较好的区域、小城镇、农村，改造和升级为产业新城或特色小镇这样的模式。

带有一定产业特性的小城镇在大城市周边的生命力会越来越强，这点在许多发达国家已经得到验证。很多大城市周边，星罗棋布地分布着以某些产业为主题的小型城镇，比如汽车城等。产业集聚，意味着人才和各种配套设施也相应集聚。我国正处在产业升级的过程中，部分产业需要转移出大城市，因此这个趋势将来会更加明显。

特色产业塑造新型小镇

在您心中，特色小镇中的"特色"意味着什么？

朱轶俊：所谓特色小镇，这个"特色"指的是它的产业特色。不同的产业会引出小镇形态和精神层面的不同特征。比如，"电影小镇"的建筑形态会更加活泼，里面的工作人员、配套设施的分布就会有别于以机械或是加工为特色的小镇。

我们现在所谈的产业，包括了第一产业、第二产业和第三产业。第一产业催生出以农业为主题的小镇，比如，观光农业、亲子农业；第二产业就是所谓的加工生产，我国目前有很多这样的小镇，比如，红木小镇、机器人小镇、医药小镇；到第三产业这个层面，有健康小镇、旅游小镇、温泉小镇……它们服务于人们更高的需求，把第一产业、第二产业的产品延伸到服务和相关领域，通过展示、服务、信息等方面获得发展。未来这个类型的小镇会越来越多。

从您的经验来看，怎样把握规划设计的"在地性"，从而赋予每个项目独特性？

朱轶俊：我们现在做的每个项目都具有在地性。一个小镇的产业特色跟规划地所在的人文环境、地理环境是分不开的。

When were you involved in town planning and design? How did you see the development trend and planning concept at that time and why?

Zhu Yijun: We started the town and country planning 8 years ago, including projects about city-industry integrations, beautiful countryside and featured towns.

Before starting those projects, we found out that the urbanization development in China has already reached saturation point. The majority land in China is still for countryside and town, which is tremendous to develop in the future. So, in the coming 5 to 10 years, the significant part of urban and rural development will be the suburban town renovation and regeneration, fellow town planning and featured town development.

When looking back this process, what changes of the planning background and tasks have taken place?

Zhu Yijun: The process is comprised of researching, practice, enhancement and redevelopment.

We were not very aware of what we were doing 8 years ago. What we were doing are just putting the components like residential and industrial park of the urban space into suburban part and adding in new components. But gradually, we found out that urban disease appears after urbanization developed in a certain stage. Air pollution causes people to desire better environment. This change leads to real requirements. The well-developed Internet and Internet of Things bring possibilities to remote office and remote production. Therewith, new industrial city and featured town come into being after the renovation and regeneration of suburban, town and countryside.

Towns with special industries are developed better in peri-urban areas, which have been proven in many developed countries. There are many industry-themed towns around mega cities, such as automobile city. The gathering of industries will lead to the gathering of talents and facilities. As China is in the process of industry upgrading, part of the industries need to be moved out from the mega cities, which will be more obvious in the future.

Specialty industries sculpture new types of towns

What does the "feature" of featured town mean to you?

Zhu Yijun: In my opinion, the feature of the towns means the industry characteristics. Different industries will lead to different types and mental conditions of the towns. For example, the architectural form of movie town will be more vivid. The staff and facilities distribution will be totally different from the towns of machinery or production.

The industries we mentioned here include primary industry, secondary industry and tertiary industry. The primary industry results in the agricultural towns, such as sight-seeing agriculture or parent-child agriculture. The secondary industry is what we called production. We have a lot of towns of this type, such as rosewood town, robot town and medicine town. Regarding to tertiary industry, we have healthcare towns, tour towns and hot spring towns, which serves the further requirements and extend the services of primary and secondary industries by exhibition, service and information. This type of towns will grow rapidly in the future.

How do you adapt local condition to planning to make each project unique?

Zhu Yijun: All the projects we are designing are adapted to the local condition. The industry characteristics of the town are unseparated from the humanistic and geographical environment.

我们针对所有项目都有一个运作的程序：设计团队会先跟投资方或是当地政府进行交流，策划团队会对本地的文化和地理元素进行挖掘。有了这些资料以后，根据我们对当地产业的评估，将产业策划和我们的规划结合在一起，才能够形成所谓的概念规划。有了概念规划，接着要把它落实到论证的层面，对建设的指标、建设的量、产业导入的可能性以及业态等进行论证。然后，这个规划才可以作为实施性的方案，才具有指导性，才可以具体开展项目的设计。

每一个规划，都为将来留下记忆

您认为什么样的城市更新项目，可以被认为是成功的？

朱轶俊：特色小镇或是美丽乡村项目，是指对城市之外（主要是农村）的地方进行重新打造，或者是在城市之外的一些片区进行产业布局和导入新城市运动。城市更新则是另外一个课题。

目前，很多城市建造于二十世七八十年代的区域，在交通功能、形态等方面已经不符合现在发展的需求了。

城市更新成功与否，主要在于对城市本身的继承。一个城市应该是有年轮的，现在很多城市的做法是全部拆掉，重新再建。这样的城市就只能代表一个时代。如果任何一个离开这个城市的人再回来的时候，都能找到自己的记忆，那么它就是一个有内容的城市。对我而言，城市更新并不意味着把过去抹掉，而是要去进行提炼、维护和提升。

目前，我国许多规划设计项目的规模都比较大，对当地的社会经济发展有着重要的意义，因此设计师与甲方之间会产生许多深入的沟通、碰撞，您能否谈谈这方面的情况？

朱轶俊：现在的规划都是在非常大的层面上进行的。20年前，我们做一个二三万平方米的项目，就算很大的，现在做到20万平方米都不算大型项目，常常一个项目就是二三百万平方米。这不是好与不好的问题，它说明我们的步子加快了。从规划师的角度来讲，我希望能够进行更大范围内的控制，去做整体规划和整体建设，这对于实现一个规划的延续性很有帮助。当然，任何规划都要与当地的经济发展和各方面条件匹配，要对它本身的文化有所继承和挖掘。

全专业的复合性团队，为规划设计夯实基础

请您谈谈思纳设计在规划设计项目方面有什么样的优势？它是怎样建立起来的？

朱轶俊：我们是一个全专业、复合性的公司，工作内容包括从前期策划到后期经济评估、运营，再到投资等部分。从专业方向来说，不仅有建筑，还有规划、室内、艺术和灯光。人员也是综合性的，有着来自世界各地的设计师。我们的优势就在于这些设计师带来了他们各自的想象力，不同的专业人员又提供了不同的思考和角度。

这些优势，是由公司本身的结构所带来的。实际上，在我2000年将思纳史密斯公司从美国引入中国之前，它就已经有150多年的历史了，它本身就是一个综合性的不动产全程服务的提供者。从另外一个角度来讲，我自己是设计师，所有项目总监、设计部门经理也都是设计师，大家对于设计领域里还没有接触过的东西都很感兴趣，对设计充满热情，大家都把项目作为自己的事业来做。

We have the same process for all the projects. The design team will communicate with the developer or government, meanwhile the strategic team will research the humanistic and geographical environment on site. And then we will start the evaluation of the local industry to get a conceptual planning. After that, the feasibility study of architectural index, quantity, industry introduction and components will be the next step. Finally, we will get the practical plan.

Every plan will become the memory in the future

What kind of urban regeneration project do you think is successful?

Zhu Yijun: Featured town or beautiful countryside project are mainly about the renovation of the areas out of the city (mainly countryside) or adding new industry planning in the suburban areas. Urban regeneration is another topic.

So far, many cities are built around 1970s or 1980s. So the transportation cannot meet the current development requirements.

The success of urban regeneration lies on the heritage of the city. Each city should have its own annual ring. Most of the urban regeneration is only to demolish all the buildings and build a new one. In this way, city can only represent for one period. If someone comes back after leaving can find his memories, then it is a city with its own content. To me, urban regeneration never means demolishing, but extracting, maintaining and improving.

Nowadays, the planning projects in China are always very large, which means a lot to local socioeconomic development. Therefore, designers may have intensive communication and brainstorming with the developer. Could you share with us some stories?

Zhu Yijun: As you know, the planning projects nowadays are quite large. A project of 20,000 to 30,000 square meters may be very large 20 years ago. But now projects of 2,000,000 to 3,000,000 square meters are very common. It has nothing to do with good or not, which only means our stride is bigger than before. As a urban planner, I prefer to bigger scale projects, which ensures integrated planning and construction. It helps a lot in the continuity of planning. And definitely, the planning should match the local economic development and other situations. It should also stick to the heritage of culture.

Comprehensive team lays solid foundation of the urban planning

What are the advantages CNASMITH Group has in planning? And how is it established?

Zhu Yijun: CNASMITH Group is a comprehensive company with talents of all professions, from preliminary planning, to economic appraisal, operation and investing. The team is comprised of architects, urban planners, interior designers, art designers as well as lighting designers. The designers come from all over the world. They bring their own imagination to their projects and different designers can provide different thinking skills and perspectives.

All these advantages are from the company structure. Before I introduced CNASMITH Group from U.S. to China in 2000, it already existed for more than 150 years . It is a comprehensive service provider for real estate. From another standpoint, all the directors and managers including me are designers. We are very interested in new things in design and passionate to finish the project.

能否结合具体项目，谈谈规划设计的流程？如何通过团队建设来实现设计品质的管理？

朱轶俊： 设计品质是通过流程的管理来进行控制的。我们有一个比较完整的流程控制，它的关键在于这样一个理念——作为设计师，要用专业的技能和艺术的眼光去实现投资人的想法，帮助投资人从技术的角度解决问题，让它达到双方都认可的程度。我们是一家商业性的公司，希望每个项目投资人都能够获得成功。我们就是基于这样的理念进行管控的。所有的服务，所有的流程都因为有这样的一个理念而能够贯彻下去。

作为设计企业的掌舵人，您如何推动设计团队在研究、设计方面的持续进步？

朱轶俊： 做设计本身是一件很辛苦的事情，所以设计公司的老板如果自己不是学设计专业的或对设计没有兴趣，他就感觉不到设计过程中的痛苦和快乐。所以要谈我们公司内部对设计的推动，就是我会具体参与到很多项目当中去，包括具体的讨论、和业主的对接、项目选址的考察，使我与设计师之间有了对话的前提，这对整个项目的推动是最实际的。

除了规划与建筑设计，您本人在艺术等其他领域的造诣也颇为深厚。这些复合的经验，对于您从事设计与企业管理有什么样的意义？

朱轶俊： 除了规划和建筑设计之外，我对艺术和收藏也很感兴趣——一是对文化记忆的收集，通过收藏各种各样的物件来进行文化记忆的收集，小到茶壶、茶杯等器皿，大到整栋房子；二是一些专项收藏，包括中国的书法、碑帖和环喜马拉雅地区艺术品的收集。在职业之外，我对于一些固定的艺术领域也有所研究。

作为设计师，个人生活的品位、鉴赏能力、对事物的理解能力实际上跟自己设计作品的好坏与深度是有关联的。如果不能很好地去理解一件艺术品，那也很难把自己的思维放到设计当中去，或者也很难理解别人设计的内涵。

当小镇成为热点，规划更要有理由

近年来，从国家到地方政府，再到开发商，都更多地将视线放在小城镇的发展上。您在跟各地政府有关部门合作这方面，有什么样的经验与心得？

朱轶俊： 这八年来，我们在做小镇或者产城融合的区域性规划设计时，除了投资商之外，最常合作的就是各地政府了。我们会说服政府把有意义的文化记忆留下，适当加入新的东西，实现经济效益和社会效益的平衡。这需要我们跟政府一起开展工作。

从我们的角度来说，首先要对这个地方有非常深入的了解，然后才能够跟政府的有关部门对话，通过对话来了解他们的需求。基于我们长期的经验和专业能力，我们也知道怎样能给投资者带来效益，包括从视觉效果到内在的产业导入，再到经济效益。最后，在投资者、当地政府和当地居民三者之间形成平衡。

为此，我们在每个项目都应用"大数据+问卷"的方法——大的策划方面使用大数据；具体细节方面使用问卷，以此了解各方的需求。这个工作方式行之有效，而且推进速度很快。我们做一个项目，从策划到规划的呈现，一般来说大概两个月就可以完成。这两个月的时间包括对项目地点的了解、大数据的分析以及就具体内容跟业主方、投资方和政府的对接，然后是规划的调整、上报、审批流程。

Could you share the planning process with us? And how do you manage the design quality by team building?

Zhu Yijun: Design quality should be controlled by process management, which we are good at. The key point is that designers should realize the investors' dreams and solve their problems by professional skills and taste in art to make a win-win. We are a business-driven company, which means we should help every developer succeed. This is how we manage and all the services and processes should be provided under this way.

As the chairman of a design firm, how do you push the steady advancement of research and design?

Zhu Yijun: To be honest, design is quite hard. And the boss of the design firm can never understand the suffering and happiness of design if he is not a designer. So I always take part in the projects, such as design charrette, communicating with the developer and site visiting. It ensures me to have common topic to talk with the team, which helps a lot during team management.

You are very professional in other fields besides planning and architectural design. How do you benefit from your rich experience during design or management?

Zhu Yijun: I am also interested in art and collection besides planning and architectural design. On one hand, I like collection about cultural memory ranging from vessels such as tea pot or tea cup to houses. On the other hand, I like special collections such as Chinese calligraphy, stone inscription rubbings and Himalaya Art. I also have studied in some certain art fields.

As a designer, lifestyle, appreciation ability and understanding skills have close relationship with the quality of design. If one can't understand an artwork, it will be very hard to blend his thought into design. At the same time, he can't understand the design from others.

After town becomes hotspot, planning should be more reasonable

In the past several years, the government and developer have pay more attention to town development. Could you share some experience in communicating with the government?

Zhu Yijun: In the past 8 years, we often cooperate with the government when we are doing the planning of towns or industry park. We will try to convince the government to keep the part with cultural memories and then add in new elements to keep the balance between economic and social benefits. It needs us to work together with the government.

From our standpoint, we should learn more about the site, which helps us to communicate with the government and get their real requirements. We always know how to bring benefits to the developer with our experience and profession, including the visual effect and industry introducing as well as economic benefits. Finally, we can find the balance point among the dev eloper, government and the local people.

As a method, we will employ massive data and questionnaire in every project. Massive data analysis helps to figure out the guideline and questionnaire can find the detail requirements. It is proven to be effective and efficient. It always takes about two months for a planning project, including the site visiting and massive data analysis. After that, we will present to the client, developer and government and finish the planning adjustment, reporting and approving.

如您所说，当您涉足特色小镇规划设计时，参与者并不多，但近年来形势有了很大的变化。对于当下这个领域里存在的"一窝蜂"现象，您有什么样的观察和判断？

朱轶俊：这是一个很可怕的现象。美丽乡村项目以居住为主，特色小镇项目则是以产业为主，如果大家一窝蜂地把后者变成城市房地产开发的延续，那就变成以产业为幌子来进行房地产开发，可能会造成特色小镇没有特色产业的问题。

还有选址的问题。许多地方本来就没有相应的风景或者产业，却被硬生生造出了所谓的美丽乡村或是特色小镇。这种做法违背了美丽乡村和特色小镇的规划建设原则，即它的在地性和基因。

为了不"迷失方向"，规划设计单位需要具体做哪些工作？

朱轶俊：作为规划设计单位，我们也需要进行阶段性的总结。目前我们有一部分产城融合的项目已经完成了，还有的处于规划状态。对这些项目进行总结，有助于发现其中存在的问题，让我们更好地考虑未来该怎么做。同时，这还能给我们的开发带来一些指导，因为我发现，从设计师转变为开发者时，理想与现实、投入与产出之间的平衡非常重要。

从设计师到投资人，换个角度"造小镇"

近年来思纳自己也参与投资，进行小镇的发展建设事业，能分享一下具体的情况吗？

朱轶俊：我们现在有两个丝绸小镇项目。一个是浙江丝绸小镇，在浙江湖州南浔的荻港。项目包含鱼塘和开发用地，共有 5.4 平方千米，其中有一个 AAAA 级旅游景区，有联合国的世界文化遗产，还将会有产业新城。还有一个是云南的文山，我们把丝绸产业引入当地，与当地的旅游业和民族风情相融合，再结合国家的精准扶贫政策，带动整个区域的发展。

您说到，产业是特色小镇项目的核心。从投资方的角度考虑，怎样利用产业的培育来进行特色小镇的培育？

朱轶俊：我们参与投资的小镇项目在产业导入方面着力特别多，比如，在我们打造的丝绸小镇里，引入了高端印刷、喷绘、礼品盒打样等产业，以及一些微产业。作为一个特色小镇，它要如何吸引人，让人觉得来这里是一件开心的事情？首先，要让住在小镇里的居民开心，所以我们进行微产业的导入。

我们成立了一个微产业学院，里面有几十位产业导师，他们会深入到每个项目里去教村民怎样提升他们的产品品质，把他们的微产业变成明星产业。我们现在还建立了一个"帮帮基金"，给予当地居民辅助和投资。这些辅助工作使原本不景气的生意变好了，店铺老板也变得开心，那么就能够带动周围的人。我们不断地去做这样的事情，去探索小镇的产业发展。如果这些工作做好了，就不需要拆掉那么多小镇，而是只做一些改造就好了。

对于荻港这个项目，我们改造的理念是使它 20 年后还是这样。我们认为，不变反而是最好的。可以想象，人们觉得一个水乡古镇有意思，不是因为它挂满红灯笼，到处都在卖蹄膀，到处都是小红旗、小喇叭；而是因为你去到那儿，发现白天人不是很多，大家都自然地生活，而不是都在忙。我们目前做的事情，就是通过微产业层面上的调整，让小镇呈现这样一种状态。

As you mentioned, few people are in this industry when you started town planning. But the situation changed a lot recently. How do you think about people rushing to this field?

Zhu Yijun: It is very terrible. Beautiful villages projects are mainly for residences and featured towns are mainly for industries. But if people rush to join this kind of development, it will become the same as real estate development. Then it might cause problems for featured towns.

Another problem is the wrong site selection. Fake beautiful villages or featured towns will be built without any natural heritage. It is totally against the principle of site selection for this kind of project.

Not to lose our bearing, what should planning firm do?

Zhu Yijun: We need to do summary during planning as planner. We have already finished some industry town projects with others during planning. The summary for finished projects can help us to pinpoint the problems and figure out how to do better. Meanwhile, it can also help to guide the development, because when I transfer from planner to developer, the balance between ideal and reality, input and output are significant.

Build the towns in a different way, from planner to developer

CNASMITH Group has invested in some towns development. Could you share with us?

Zhu Yijun: We have two silk town projects. The first one is Digang Village of Nanxun, Huzhou, which is a silk town in Zhejiang. This project is 5.4 square kilometers, including pounds, one AAAA national tourist attraction, World Cultural Heritage of United Nations and new industry town. Another project is Wenshan silk town in Yunnan. We introduce silk industry into this town and blend it with the local tourism and ethnic folklore to promote the development of the whole region.

As you mentioned, industry is the core of featured towns. How will the town breeding benefit from industry breeding?

Zhu Yijun: We have put a lot of efforts in industry introducing. For example, we have introduced high-end printing, spray paint and gift box printing into the silk town we invested. We also employed mini-size industry. When we are thinking how to make the town attractive, the satisfaction of the native people will be the first and foremost.

We established an institute for mini-size industry including dozens of industry tutors. Tutors will help the native people with how to promote their products and change them into star industries. At the same time, we also established "Bangbang Fund" to offer help and financial aid to the local people. We keep on helping them and explore how to help them better. And finally, we find out that there is no need for so much demolishing. Renovation works are better for the towns.

For Digang silk town project, we are devoted to make it seems the same 20 years later. In our opinion, no changes will be the best way for the town. People will never like an ancient town because of stores and crowds everywhere. The best status of an ancient town should be that people there are enjoying their lives. That is what we are involved in and devoted to.

朱轶俊 Zhu Yijun

上海思纳建筑规划设计股份有限公司执行总裁、艺术总监、高级建筑师

朱先生在项目统筹规划，城市规划设计，酒店业、展览业及零售业的设计等各种项目中都具有丰富的经验。从设计师到施工单位、材料设备供应商、开发商、高级项目管理再回到设计师的行列，他的经验在实践中得到融合，使其从容应对大型旅游和商业综合体以及城区的综合开发计划等项目并在其中获得成功的乐趣。

对项目运作的精打细算以及对艺术追求的超脱回归，让朱先生在两种截然不同的生活状态和大量的艺术收藏中不时地获得灵感。朱先生至今仍然不断地在他的工作上努力钻研并通过自己的快乐设计哲学力争达到更高的境界。

代表作品：
上海兰馨剧院
杭州凯悦酒店室内设计
北京兰海洋华纳影城
浙江嘉兴华庭街综合商业项目
世纪大道 SN3-1 办公楼地块
温岭东方不夜城
上海曙光医院迁建工程
上海颛桥老镇区总体改建规划设计
杭州解放路整体城市规划设计
上海紫园
上海实业朱家角高级居住区开发
沈阳中海国际社区
远洋沈阳科技城
庐山温泉度假酒店
上海证大喜马拉雅中心

Executive president of Shanghai CNASMITH Architectural & Planning Design Company, artistic director, and Senior architect, Mr. Zhu has rich experiences in project planning, urban planning and design, and he has accomplished a variety of projects such as the design of the hotel industry, exhibition industry and retail trade. From a designer to participating in the construction units, to be a supplier of materials and equipment, to a developer of real estate, to a senior project management, and back to be a designer again. All the experiences enable him to be handy in dealing with planning and design projects like large-scale tourism and commercial complex, and comprehensive development of urban areas.

Mr. Zhu is also an art collector, who has been able to get inspiration from the two poles of living conditions, the meticulous planning of the project and the transcendence of the art world.

Representative works:
Shanghai Lanxin Theatre
Interior Design of Hyatt Regency Hangzhou
Beijing Ocean Warner Studios
Integrated commercial project of Huating Street, Jiaxing, Zhejiang
Century Avenue SN3-1 Office Block
Wenling Oriental Sleepless City
Shanghai Dawn Hospital Relocation Project
The Overall Reconstruction Planning and Design of the Old Town of Zhuanqiao, Shanghai
Overall Urban Planning and Design of Jiefang Road, Hangzhou
Shanghai Ziyuan
Development of Zhujiajiao Senior Residential District, Shanghai Industrial Investment (Holdings)Co., Ltd.
China Overseas Property International Community, Shenyang
Sino-Ocean Science and Technology City, Shenyang
Mount Lu Hot Spring Resort
Shanghai Zendai Himalayas Hotel

CONTENTS 目录

002 THE OVERALL DESIGN OF THE BEIDAIHE NEW DISTRICT
北戴河新区总体设计

008 THE RENEWAL DESIGN OF THE OLD TOWN IN DACHANG COUNTY
大厂县老城更新设计

014 THE PLANNING AND DESIGN OF THE NEW HSR CITY IN SEMBILAN
森美兰高铁新城规划设计

020 THE URBAN DESIGN OF HANGBU AREA, SHUCHENG COUNTY
舒城县杭埠地块城市设计

024 THE PLANNING AND DESIGN OF THE NEW INDUSTRIAL CITY OF MANESAR AND SOHNA
马纳萨与索赫纳产业新城规划设计

030 THE PLANNING AND DESIGN OF THE NEW NANXUN HIGH-SPEED RAILWAY CITY
南浔高铁新城规划设计

036 THE DESIGN OF CORE DISTRICT OF FEATURED TOWN IN BEISAN COUNTY
北三县特色小镇核心区风貌设计

040 THE PLANNING AND DESIGN OF SINO-GERMAN NEW INDUSTRIAL CITY IN TIEXI DISTRICT
铁西中德产业新城规划设计

046 THE PLANNING AND DESIGN OF KANGYANG PARK TOWN IN DACHANG COUNTY
大厂县康养公园小镇规划设计

050 THE PLANNING AND DESIGN OF THE FEATURED TOWN —TWO VALLEYS AND ONE VILLAGE
两谷一村特色小镇规划设计

054 THE PLANNING AND DESIGN OF THE FEATURED TOWN OF DAYU MOUNTAIN
大禹山特色小镇规划设计

THE PLANNING AND DESIGN OF THE NEW INDUSTRIAL TOWN IN DACHANG COUNTY 大厂县产业新城规划设计	**058**	
	064	THE PLANNING AND DESIGN OF THE NEW MANRONG INDUSTRIAL TOWN 满融产业新城规划设计
THE PLANNING AND DESIGN OF THE NEW BINHUAI TOWN IN LISHUI COUNTY 溧水县滨淮新城规划设计	**070**	
	078	THE PLANNING AND DESIGN OF THE NEW INDUSTRIAL TOWN IN HUAILAI COUNTY 怀来县产业新城规划设计
THE PLANNING AND DESIGN OF THE FILM AND TV TOWN IN DACHANG COUNTY 大厂县影视小镇规划设计	**084**	
	090	THE PLANNING AND DESIGN OF THE EMERGING INDUSTRY NEW TOWN IN GU'AN COUNTY 固安县新兴产业新城规划设计
THE PLANNING AND DESIGN OF THE NEW INDUSTRIAL TOWN IN WEN'AN COUNTY 文安县产业新城规划设计	**094**	
	098	THE PLANNING AND DESIGN OF THE NEW INDUSTRIAL TOWN IN THE NORTH OF THE GRAND CANAL 大运河北产业新城规划设计
THE PLANNING AND DESIGN OF THE CORE DISTRICT OF NEW INDUSTRIAL TOWN IN JIASHAN COUNTY 嘉善县产业新城核心区规划设计	**104**	
	110	THE PLANNING AND DESIGN OF THE PEACOCK CITY IN WESTERN GU'AN COUNTY 固安县西部孔雀城规划设计
THE PLANNING AND DESIGN OF THE BEAUTIFUL VILLAGE OF LIRANGDIAN IN GU'AN COUNTY 固安县礼让店乡美丽乡村规划设计	**114**	

Page	English	Chinese
118	THE PLANNING AND DESIGN OF THE NEW YUHONG INDUSTRIAL TOWN	于洪产业新城规划设计
122	THE PLANNING AND DESIGN OF THE NEW TOWN IN NORTHERN GU'AN COUNTY	固安县北部新城规划设计
126	THE PLANNING AND DESIGN OF THE NEW HUAXIA TOWN IN YONGQING COUNTY	永清县华夏新城规划设计
130	THE PLANNING AND DESIGN OF THE NEW TOWN IN GU'AN COUNTY	固安县新城规划设计
136	THE PLANNING AND DESIGN OF THE NEW AIRPORT ELITE CITY	新空港精英城市规划设计
142	THE PLANNING AND DESIGN OF THE NEW SOUTHERN TOWN IN GU'AN COUNTY	固安县南部新城规划设计
148	THE PLANNING AND DESIGN OF THE NEW TOURIST TOWN OF GOLDEN COAST	金渤海岸旅游新城规划设计
156	THE PLANNING AND DESIGN OF THE TOURIST TOWN OF JINYUE GULF IN LINGSHUI	陵水金月湾旅游小镇规划设计
160	THE PLANNING AND DESIGN OF THE WEST EUROPEAN CULTURAL TOWN IN NINE-LOONG CANYON, HENGDIAN TOWN	横店镇九龙大峡谷西欧风情小镇规划设计
164	THE PLANNING AND DESIGN OF THE INTERNATIONAL JEWELRY CULTURE INDUSTRIAL TOWN	国际珠宝文化产业小镇规划设计
170	THE PLANNING AND DESIGN OF THE NEW SOUTHEAST ASIAN AND SOUTHERN ASIAN CULTURE AND COMMERCIAL TOWN IN DONGBA	北京市东坝东南亚、南亚文化商贸新城规划设计
176	THE PLANNING AND DESIGN OF THE WATERY TOWN OF SOUTHERN CHINA, HENGDIAN TOWN	横店镇江南水乡风情小镇规划设计
180	THE PLANNING AND DESIGN OF THE NEW TOURIST TOWN WITH THE HISTORY AND CULTURE OF THE THREE KINGDOMS	三国历史文化旅游新城规划设计

Page	English	Chinese
186	THE PLANNING AND DESIGN OF THE INTERNATIONAL TOURISM ISLET IN TANGSHAN BAY	唐山湾国际旅游岛规划设计
190	THE DESIGN OF XINGDIAN CITY GROUP, THE TOWNSHIP OF WAGANG, QUESHAN COUNTY	确山县瓦岗镇镇区邢店组团城市设计
196	THE RENEWAL DESIGN OF QUJING DOWNTOWN	曲靖市中心城区更新设计
200	THE PLANNING AND DESIGN OF THE NEW TOWN OF DONGHU	东湖新城规划设计
206	THE PLANNING AND DESIGN OF THE CREATIVE INDUSTRY TOWN IN DALI CITY	大理创意产业新城规划设计
210	THE PLANNING AND DESIGN OF DONGJIN INTERNATIONAL TOWN	东津国际小镇规划设计
216	THE PLANNING AND DESIGN OF THE SOUTH OF UNIVERSITY ROAD	学府路南侧新城规划设计
220	THE PLANNING AND DESIGN OF YUDU GONGJIANG RIVER WATERFRONT NEW CITY	于都贡江滨水新城规划设计
226	THE PLANNING AND DESIGN OF NEW LAISHUI COUNTY	涞水县新城规划设计
230	THE PLANNING AND DESIGN OF THE NEW INTERNATIONAL TOWN OF TOURISM VACATION IN BAIYANGDIAN	白洋淀国际旅游度假新城规划设计
236	THE PLANNING AND DESIGN OF SHANGHAI POLAR SEA WORLD	上海极地海洋世界规划设计
240	THE PLANNING AND DESIGN OF THE NEW HIGH-TECH AREA, ZHUOZHOU	涿州高新技术产业新城规划设计

THE OVERALL DESIGN OF THE BEIDAIHE NEW DISTRICT

北戴河新区总体设计

Location	Area	Time
河北省秦皇岛市 Qinhuangdao City, Hebei Province	425 平方千米 425 square kilometers	2016 年 2016

北戴河新区凭借毗邻首都的区位优势和生态环境资源优势，已被列为深化改革和创新发展的国家级新区。在京津冀协同发展上升为国家战略的背景下，北戴河新区将进一步成为首都教育、医疗等非核心职能向外疏解的主要承载区之一，并逐步成为京津冀新发展格局中的重要战略支点。但是由于北戴河新区前期开发中已经出现了环境品质不高、过度消费滨海资源、滨海天际线无序发展等问题；如何发挥新区自身优势，打造新区高品质的城市环境，塑造符合地域特征的城市风貌，引导城市有序发展，已成为新区规划建设工作的重中之重。

工作目标

研究城市发展现状及存在的问题，梳理城市资源特征，确定城市形象发展目标，制定总体城市形态引导策略，引导城市有序发展，塑造符合地域特征的城市风貌。

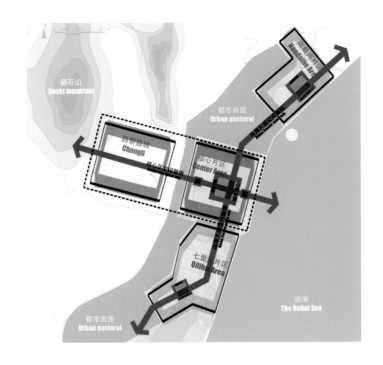

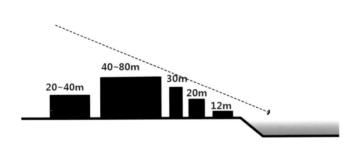

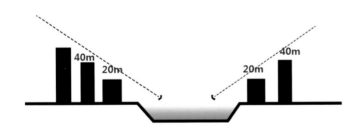

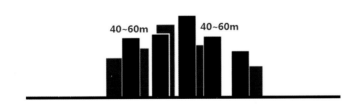

The Beidaihe New District with its geographical advantage of being adjacent to the Beijing city and its natural advantage of ecological environment has been listed as a new district for deepening reform and innovational development on the national level. Against the background of a coordinated development of the Beidaihe New District of Beijing, Tianjin and Hebei province, the Beidaihe New District will step forward to become one of the major receptive districts for the capital noncore functions such as education, medicare to release its pressure, and gradually form a major strategic point in the new economic development pattern for Beijing, Tianjin and Hebei province. But some problems have emerged in the preliminary development of the Beidaihe New District, such as low environment quality, the excessive consumption of the coastal resources and unsustainable development of coastal skyline. It's of vital significance to give a full play to its own advantages, to create high quality environment of the new district, and to forge an urban landscape that is in line with its geographical characteristics. It has emerged as a priority to lead an orderly development in planning and construction of the Beidaihe New District.

OBJECTIVES

It is to study the current status of urban development as well as the existing problems, to sort out the characteristics of urban resources, to determine the development goal of city image, to formulate the guidance strategy of the city overall framework, to lead the orderly urban development, and to shape the city outlook that meets its geographical characteristics.

总体定位

设计将形象定位为"青山绿野，滨海新田园"。设计将功能定位为国际知名的高端滨海度假旅游胜地、面向全球的健康产业集聚区、精英汇聚的创新创业示范引领区域。

设计概念与策略

设计概念：带状城市，组团发展；一体两翼，山海田园；双面城市，风貌多元；绿道水网，连而不粘；点面轴心，多维缝合。

设计策略：依田伴海，结合资源特征控制总体形态；显融露透，结合自然基底打造城市形象；多元整合，结合功能布局塑造片区特色；疏密有致，结合生态保护制定开发策略。

总体形态控制

规划由城市核心区和昌黎县构成城市发展的统一体，由南戴河片区和七里海的会议及健康片区构成城市的两翼。由此，形成一体两翼的城市格局，与山海田园相融。

整个城市沿滨海新大道呈带状发展，形成三大组团，每个大组团又强调小的功能组团的划分。

OVERALL ORIENTATION

The image of the new district is orientated as "new coastal fields with trees and green mountains". It will function as international famous high-end coastal holiday resort, an accumulation area of health industry facing the world, and an innovative, entrepreneurial model area as an agglomeration of elites.

DESIGNING CONCEPT AND STRATEGY

Design concept:It is going to be a strip city developed in groups, which has one main body and two wings with mountainou, sea, field and garden scenery; the city is supposed to be double-sided with multi styles. It is also equipped with green belts and water system, connected to each other but not clinging. All its points and spaces, axes and cores bind with each other flawlessly.

Design strategy:With both the agricultural fields and the sea close by, the design combines the characteristics of natural resources to control the general pattern. Highlighting, integrating or revealing, it combines the natural scenes to make the image of the city. Coordinating multi elements, it combines all the functions to design the layout and create unique features of different areas. With proper density and scarcity, it combines ecological conservation to formulate development strategies.

CONTROL OF THE OVERALL PATTERN

The city core area and Changli county constitute a unified body of city development. The two wings refer to Nandaihe district and Qihaili conference and health areas. Thus, there is to be a city pattern of one main body with two wings, which fits into the mountain, sea, field and garden.

The whole city develops in stripe type along the new coastal boulevard, forming three big groups, in which each emphasizes the division of smaller function groups.

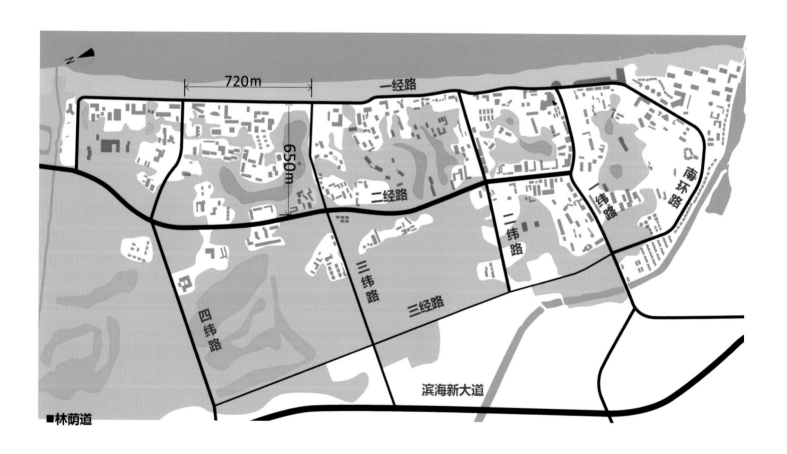

■林荫道

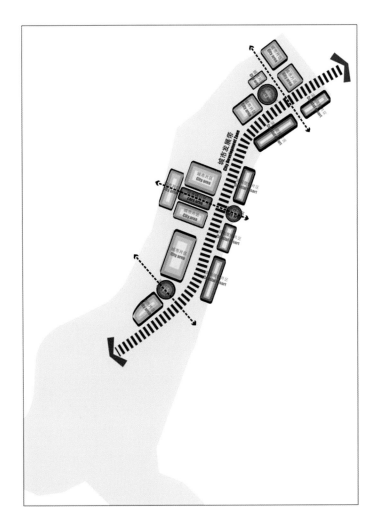
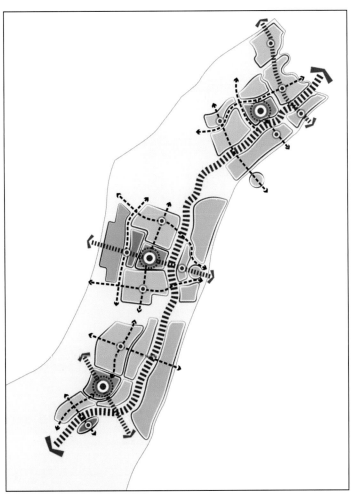
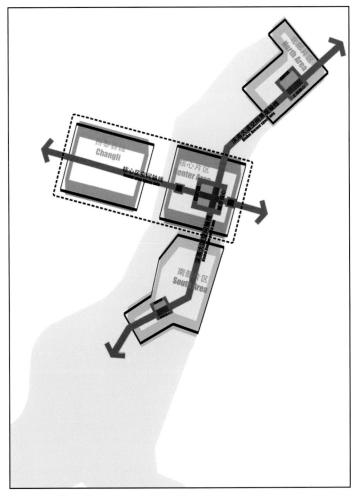

THE RENEWAL DESIGN OF THE OLD TOWN IN DACHANG COUNTY

大厂县老城更新设计

Location
河北省廊坊市
Langfang City, Hebei Province

Area
1.27 平方千米
1.27 square kilometers

Time
2016 年至今
Since 2016

依托现有的公共设施，增加新的功能空间以使新旧联动，给整个片区带来新的面貌和充沛的活力。

方案1：双核驱动、两轴连贯、片区融合

双核即综合服务核心与文化核心。综合服务核心是指依托现有游泳馆、社区商业、养老设施等公共功能组团，增加新的服务功能，形成综合服务核心。文化核心是指结合现有的文化广场和评剧团等文化产业组团，形成片区的文化核心。

生活活动轴线沿北新东街整合现有的开放空间、沿街商业空间，规划综合服务核心，增加社区公园、道路绿化等开放空间，形成主要生活的活动轴线。同时，沿大安东街结合现有的行政单位、沿街商业空间等，改造和提升沿街界面，增加活力功能空间，形成行政商业综合服务轴线。

此外，依据规划路网，结合功能分布形成不同片区，再结合现有的公共地块，提升商业功能品质并增加新的公共功能，形成片区服务中心，实现片区融合。

This project is to rely on the existing public facilities to increase the new functional space, integrating the old and new, to bring a new look and vitality throughout the area.

PROGRAM 1: DUAL CORES DRIVES, TWO AXES INTEGRATION, AREA INTEGRATION

The dual cores are the comprehensive service core and cultural core. Comprehensive service core is to rely on the public function group such as the existing swimming pool, business community, pension facilities etc., to add new service functions, in order to form a comprehensive service core. Cultural core means to combine cultural industry group such as the existing cultural square, and Pingju opera troupe, in order to form the cultural core.

The living and activity axis, along the Beixindong Street, is to integrate the existing open space and commercial space along the street, to plan the comprehensive service core, to increase open space by adding community parks, road greening, and to form action axis of the main living activities. At the same time, along Da'andong Street, it is to combine the existing administrative units, commercial space along the street, to transform and upgrade the interface along the street, to increase the vitality function space, and to form a comprehensive service axis of administration and commerce.

In addition, according to the planned road network, it is to form different areas according to the distribution of functions; in combination with the existing public land, it is to improve the quality of the commercial functions and increase new public functions, to form area service centers, and to achieve regional integration.

01.评剧团
02.文化广场
03.人民法院
04.文化产业园
05.体育馆
06.御园小区
07.世纪花园
08.公路管理站
09.社区公园
10.魅力中心
11.养老院
12.中石油
13.城管小学
14.幼儿园
15.世纪家园
16.福华庄园
17.嘉怡丽都
18.休闲步行商业街
19.水务局
20.工商管理局
21.供电局
22.建设银行
23.丰泽家园
24.传统步行商业街
25.县政府
26.市民广场
27.德仁吉第小区

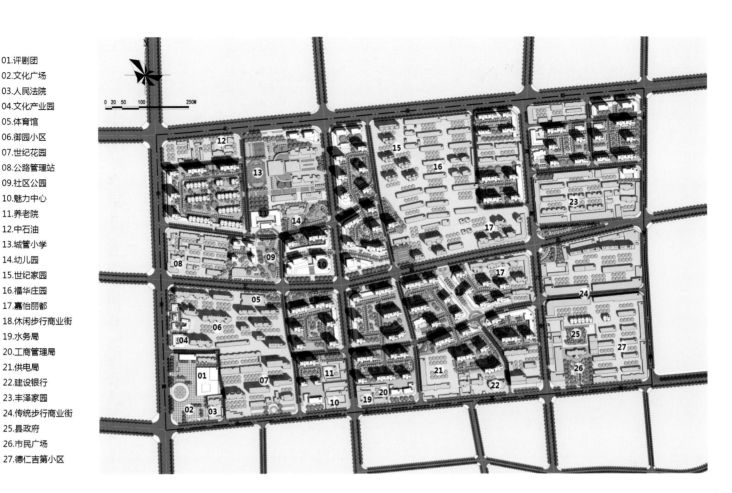

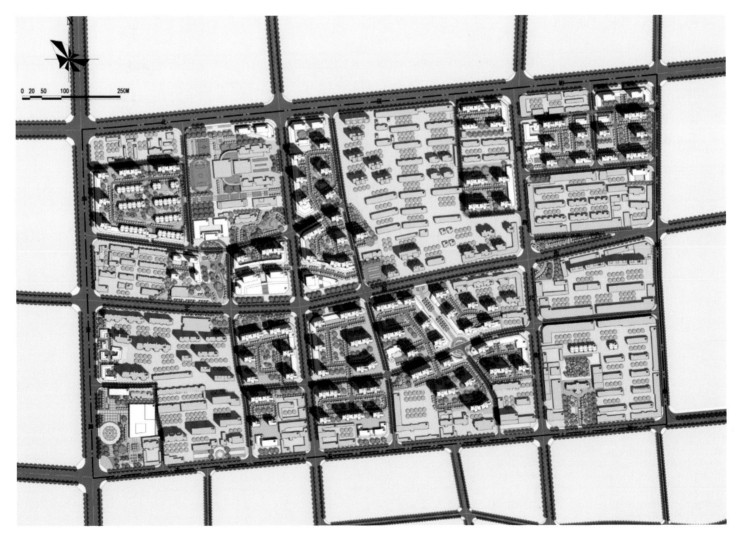

方案2："两心两轴九片"

"两心"：结合现有体育馆与公共服务设施，增加新的城市公共服务功能以形成综合服务核心，并结合现有文化广场和评剧团，增加文化产业组团以形成文化核心。

"两轴"：一是沿北新东街整合现有的开放空间和沿街商业空间，规划综合服务核心，增加社区公园、道路绿化形成主要生活的活动轴线；二是沿大安东街结合现有行政单位、沿街商业空间，改造和提升沿街界面，并增加活力功能空间，形成行政商业综合服务轴线。

"九片"：指依据规划路网，结合功能分布形成九个片区。它们又分为核心片区、文化片区、居住片区三种类型。

PROGRAM 2: "TWO CORES, TWO AXES AND NINE AREAS"

"Two cores": It is to combine the existing stadium and public service facilities, to increase the new city public service function to form a comprehensive service core. And combining the existing cultural square and Pingju opera troupe, it is to increase the cultural industry group to form a cultural core.

"Two axes": The first is, along the Beixindong Street, to integrate the existing open space and commercial space along the street, to plan comprehensive service core, to increase community parks, road greening to form the action axis of the main living activities. The second is along the Da'andong Street, it is to combine the existing administrative units, commercial space along the street, to transform and upgrade the street interface, and increase the vitality function space, to form a comprehensive service axis of administration and commerce.

"Nine areas": It is to form nine areas according to the planning of the road network, and combing functional distribution. The nine areas are divided into core areas, cultural area and living area.

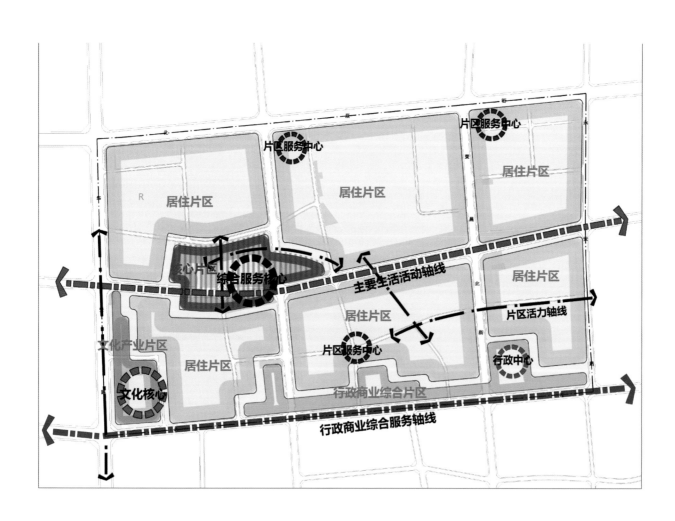

01. 评剧团
02. 文化广场
03. 人民法院
04. 文化产业园
05. 体育馆
06. 御园小区
07. 世纪花园
08. 公路管理站
09. 社区公园
10. 魅力中心
11. 养老院
12. 中石油
13. 城管小学
14. 幼儿园
15. 世纪家园
16. 福华庄园
17. 嘉怡丽郡
18. 休闲步行商业街
19. 水务局
20. 工商管理局
21. 供电局
22. 建设银行
23. 丰泽家园
24. 传统步行商业街
25. 县政府
26. 市民广场
27. 德仁吉第小区

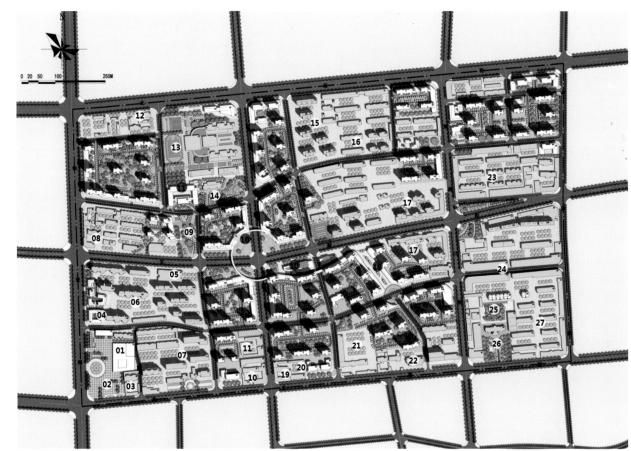

THE PLANNING AND DESIGN OF THE NEW HSR CITY IN SEMBILAN

森美兰高铁新城规划设计

Location

马来西亚
Malaysia

Area

34.5 平方千米
34.5 square kilometers

Time

2016 年至今
Since 2016

本项目靠近马来西亚芙蓉市，周边有机场、生物制药高端产业、科技谷、综合开发区、教育等设施。

"一核三轴十片区"

"一核"：由中央公园整合芙蓉高铁站商务区、世贸岛和智慧科技港共同构成新城的复合型中央商务区（CBD）。

"三轴十片区"：由山地公园、体育公园、高尔夫公园、中央湖景公园等主要公共开放空间构成城市生态轴，并延伸渗透至各个城市功能板块；由两条景观大道构成的T形城市功能轴，联系区域主要出口及公共配套设施。这三条轴线形成城市空间骨架，整合起城市的产业和生活片区，营造出最富魅力的工作和生活环境。

This project is close to Seremban city, Malaysia, which is in adjacent to airport, high-end biopharmaceutical industry, high-tech valley, comprehensive development zone and education, etc.

"ONE CORE, THREE AXES AND TEN SECTIONS"

"One core" in the planned program is the comprehensive CBD in the new city made up of the central park's integrating the business area of Seremban HSR station, world trade island and smart technology port.

"Three axes and ten sections" mean the urban ecological axis composed of public open space including mountainous park, sports park, golf park, and central lake scene park, and then penetrating every city function block. The T-shaped city functional axes consisted of two boulevards are to connect the main exits and public supporting facilities within this area. These three axes form the framework of the city, integrating principal industry and living sections, so as to create the most attractive environment for working and living.

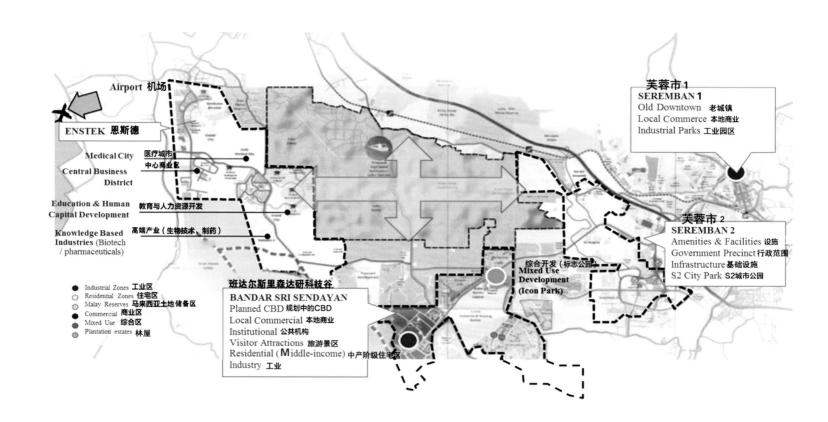

THE URBAN DESIGN OF HANGBU AREA, SHUCHENG COUNTY

舒城县杭埠地块城市设计

Location	Area	Time
安徽省六安市 lu'an City, Anhui Province	约 44.5 平方千米 about 44.5 square kilometers	2016 年至今 Since 2016

舒城县杭埠产业新城依托合肥环巢湖战略，发展由圈层向南、环湖转变，将对周围城镇产生深远影响。从地理位置上看，杭埠是舒城在空间上与合肥关系最为紧密、最具发展潜力的地区，规划总面积约44.5平方千米，是舒城对接合肥环巢湖战略的切入点。

项目格局基本上分为东、中、西三个大的板块：东片区为未来城市发展的主要片区；中片区为以产业为主的产城融合区；西片区为老镇改造和产业提升发展的综合功能区。

"产 + 城"的发展格局

利用杭埠产业新城临近合肥环巢湖的区位优势，建立"三河 + 杭埠"中心，引领舒城的发展。

在产业协作方面，积极拓展"杭埠产业，服务合肥"的发展思路，与三大产业新城错位发展，建立自身特色产业链，打造环巢湖地区重要产业增长极。

在文旅景观方面，通过整体考虑杭埠与三河的竞合关系，借势开发旅游服务和休闲度假项目的，提升三河的旅游服务品质，共同打造环巢湖滨水文化旅游重要节点。

New Hangbu industrial city in Shucheng County, based on the strategy of Chaohu rim area, Hefei, will exert a profound impact on the surrounding towns, transforming its development from circles and layers to the south and around the lake. Geographically, Hangbu in Shucheng is the closest to Hefei, and thus has the greatest developmental potential. The planned area is about 44.5 square kilometers, and marks a starting point for Shucheng to be docked to the strategy of Chaohu rim area, Hefei.

The project pattern is currently divided into three sections: the east, the central and the west. The east section is the main area for further urban development, the central section is an integration of industry and city, with industry as its core, and the west section is to be a comprehensive functional area for the transformation of the old town and industry upgrading and development.

THE DEVELOPMENT PATTERN OF "INDUSTRY + CITY"

Taking the regional advantage of new Hangbu industrial city' closeness to Chaohu rim area, Hefei, it is to build a "Sanhe + Hangbu" center, so as to lead the development of Shucheng county.

In industrial cooperation, it is to actively spark the development idea of "Hangbu industry to serve Hefei", and achieve complementary development with the other three new major industry cities, to establish its own characteristic industry chain, and to create an important industrial growth pole in the Chaohu rim area.

In culture and tourism, it is to take into comprehensive consideration of the competition and cooperation relationship of Hangbu and Sanhe, to develop tourism service and leisure and vacation project, and to promote the quality of tourism service in Sanhe, and to co-build a key node for Chaohu rim waterfront culture tourism.

总体规划:"一核两轴六心七片"

规划形成了"一核两轴六心七片"的结构,包括一个城市商务核心、两条城市发展轴、三个城市服务核心、三个生态核心和七个功能片区。

THE OVERALL PLAN: "ONE CORE, TWO AXES, SIX CORES AND SEVEN BLOCKS"

The program constitutes a structure of "one core, two axes, six cores and seven blocks", including one urban business core, two city development axes, three urban service cores, three ecological cores and seven functional blocks.

THE PLANNING AND DESIGN OF THE NEW INDUSTRIAL CITY OF MANESAR AND SOHNA

马纳萨与索赫纳产业新城规划设计

Location

印度新德里南部
South of New Delhi, India

Area

A 地块占地 3.44 平方千米，B 地块占地 6.07 平方千米
Plot A covering an area of 3.44 square kilometers, Plot B covering an area of 6.07 square kilometers

Time

2016 年至今
Since 2016

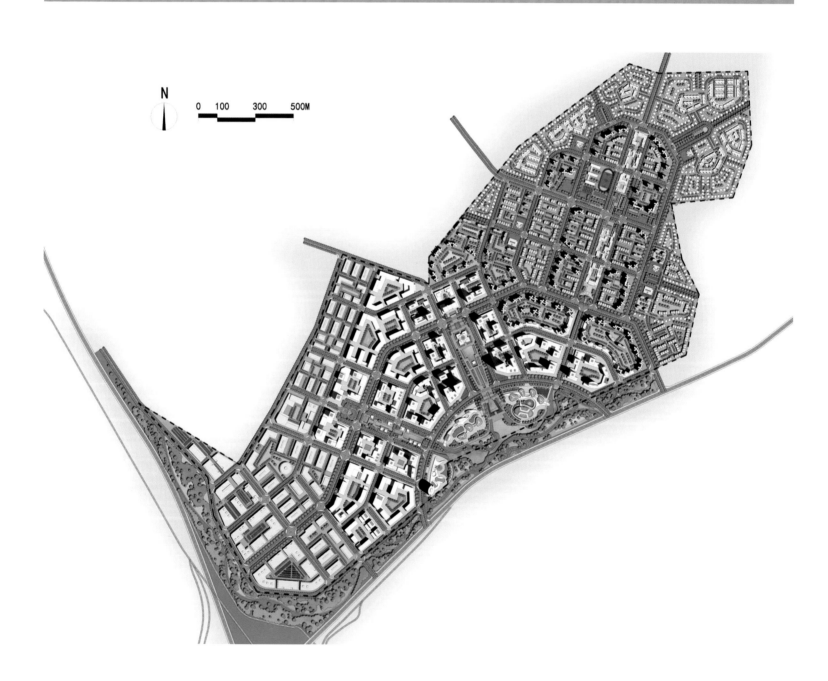

地块 A：马纳萨——共创印度产城融合新标杆

规划将打造出一个创智产业高地和生态乐居典范，融合商务服务、产业创新、文化传媒与绿色生活，使之发展成印度产城融合的新标杆。

方案提出"绿核居中、三心辉映"，在基地中部打造文化休闲生态核心，向外辐射形成产业服务中心、商务中心、城市生活中心，引领片区融合发展；同时实现轴线延展、廊道渗透，即以"绿核"为区域引擎，向外辐射产业发展轴、中央商务轴、乐居服务轴，打造区域重点界面空间，塑造城市多元形象。方案还引入城市生态"绿环"提升环境品质，形成"城绿"相依的空间骨架。

PLOT A: MANESAR—CREATING A NEW BENCHMARK OF INDIAN CITY-INDUSTRY INTEGRATION

The plan is to build a model of innovative industrial highland and eco-friendly residence, integrating business services, industrial innovation, cultural media and green life, and to make it a new benchmark of India city-industry integration.

The project advocates "centralizing the green core and three centers affecting each other". It is planned to build a cultural and leisure ecological core in the center of the base, which radiates externally to form industrial service center, business center and city residence center accordingly, promoting the integrating and development of the areas. Meanwhile, the design is to extend in axes and permeate through corridors. Namely, it is to use the "green core" as regional engine, radiating industrial development axis, central business axis, and enjoyable residence service axis, to establish regional key interface space, and to shape the plural images of the city. Urban ecological "green ring" is also introduced to enhance the quality of the environment and form an interdependence spatial framework of the city and the green ring.

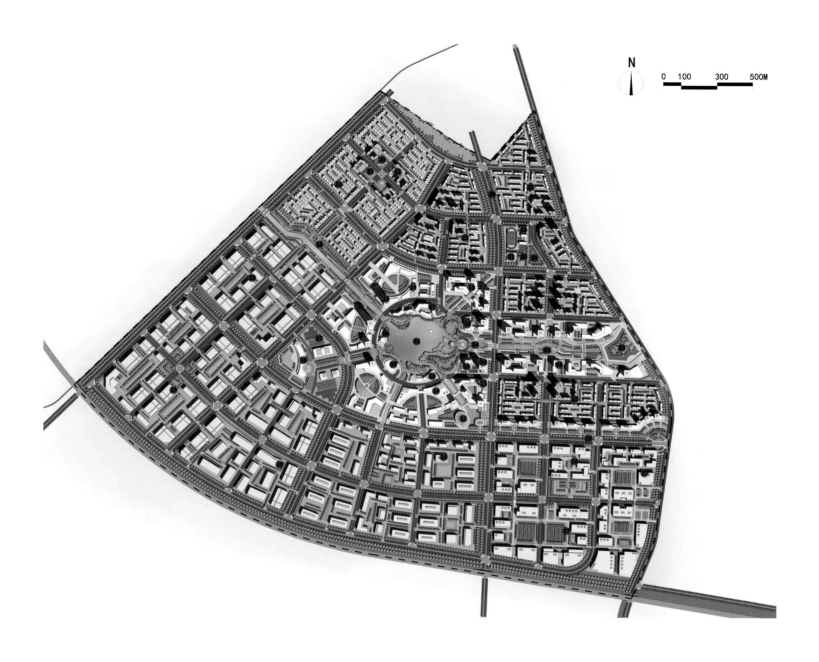

地块 B：索赫纳——"一核一环三轴"的布局

规划形成了如下布局：

"一核"：由中央公园、商务中心、科研中心、文体场馆共同构成复合型核心区。

"一环"：生态绿环串联各个城市功能。

"三轴"：由中央公园、商业中心、综合办公楼、体育公园、科研中心构成的城市生态功能轴；由两条景观大道构成的十字形城市功能轴，联系区域主要出口及公共配套设施。三条轴线形成城市空间骨架，整合起城市的五个主要产业和生活片区，营造出最富魅力的工作和生活环境。

PLOT B: SOHNA—THE LAYOUT OF "ONE CORE, ONE RING AND THREE AXES"

The project comes to the layout as the following:

"One core": a complex core area consist of a central park, a business center, a scientific research center and a venue for culture and sports.

"One ring": an ecological green ring links all the urban functions.

"Three axes": the city eco function axis consist of a central park, a business center, a comprehensive business office building, a sports park, and a scientific research center. A cross-shaped city function axis is formed by two landscape avenues, which connects the principal exits and public facilities. The three axes constitute the spatial framework of the city, integrating the city's five major industrial and living sections, and creating the most attractive working and living environment.

THE PLANNING AND DESIGN OF THE NEW NANXUN HIGH-SPEED RAILWAY CITY

南浔高铁新城规划设计

Location

浙江省湖州市
Huzhou City, Zhejiang Province

Area

23.54 平方千米
23.54 square kilometers

Time

2016 年至今
Since 2016

| CENTRAL BUSINESS
中心商务 | INTELLIGENCE INDUSTRY
智慧产业 | ECO LEISURE
生态休闲 | HABITABLE
绿色宜居 |

   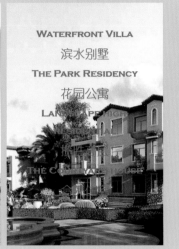

北片区：沪浙创新高地，世界的"江南客厅"

把握南浔大道、浔练公路与新318国道的"金十字"城市核心，创建公园岛链，使滨水公园带成为全城生态基础。

明确"东城西产"的格局，东部为城市生活片区，西部为产业片区。

实施"水绕岛居"，使滨水景观带渗透全城，实现城市片区组团化。

THE NORTHERN AREA: INNOVATIVE HIGHLAND OF SHANGHAI AND ZHEJIANG, "DRAWING-ROOM IN THE SOUTH OF THE YANGTZE RIVER" OF THE WORLD

It is to make use of the "golden cross" city core of Nanxun Avenue, Xunlian Highway and the new 318 National Road, to create a park-island chain and to built the entire city's ecological base by constructing a riverside park zone.

It is to stipulate the setup of "city in the east and industry in the west", meaning that the urban residence area is in the east and manufacturing area is in the west.

"Living areas surrounded by water" is to be carried out, so that the waterfront scenery zone can permeate the whole city, and make different city areas group together.

南片区：以绿环串联四个主题小镇

一心四园多组团，

绿楔渗透聚绿环。

运河文化休闲核，

度假双养两相宜。

规划在区内打造休闲绿环以串联四个主题小镇。

THE SOUTHERN AREA: GREEN ZONE WILL CONNECT FOUR THEME TOWN IN SERIES

One core with four gardens and many groups,

The green belt permeates the city and froms a green zone.

The Grand Canal culture is fully employed for leisure.

Vacation and elderly care are both excellent.

The plan is to create four theme towns with a leisure green zone connected in series.

THE DESIGN OF CORE DISTRICT OF FEATURED TOWN IN BEISAN COUNTY

北三县特色小镇核心区风貌设计

Location	**Area**	**Time**
河北省廊坊市 Langfang City, Hebei Province	20.57 平方千米 20.57 square kilometers	2015 年至今 Since 2015

项目旨在建设更高品质的城市空间，融入现代国际都市的休闲服务配套，营造更具交往性、归属感和吸引力的城市生活环境，打造独一无二的生态水韵之城。

构建以人为本的城市空间体系

为营建具有持续活力和吸引力的魅力之城，方案提出四大策略：

重塑城市内涵——新背景下的城市定位提升与确定；

优化空间结构——通过社区结构的组织、公共服务设施的再分配和绿地开放空间系统的梳理，结合水元素将城市空间结构进一步整合、梳理和优化；

控制与引导——制定控制原则和标准，指导城市整体开发建设；

修补现有问题——提升已建成区域空间品质和修补现有的问题。

"三轴二核，五区合环，水绿交融，生态共生"

方案通过水网和绿地引入潮白河景观资源，让二者融入城市风景，重塑城市格局。本次规划结构如下：

"三轴"：三条城市发展轴线。方案依托两条景观大道（迎宾大道、蒋谭路）和一条城市轴线大道（夏祁路）形成三条城市发展轴。

"二核"：两个城市核心。即一个中央商务核心区（CBD）和一个休闲商务核心区（RBD）。

"五区"：五大生态居住片区。围绕城市核心，依托绿轴、绿环形成五大居住片区。

The project aims to build a higher-quality urban space, integrating leisure-service facilities of modern international metropolis, to create a city life with more interaction, more sense of belonging, more attraction, and to build a unique city with ecological charm of water.

CONSTRUCTING A HUMAN - ORIENTED URBAN SPATIAL SYSTEM

To build a city with sustained vitality and attraction, the program proposes four strategies:

Reshaping the city connotation—enhancing and specifying the city orientation under the new background;

Optimizing the spatial structure—this is to further integrate, sort and optimize the urban spatial structure through organization of community structure, redistribution of public service facilities, sort of green open space system, and together with water element, Control and guidance—formulating control principles and standards to guide the overall development and construction of the city;

Solving the existing problems—it will improve the quality of the built region and solve the existing problems.

"THREE AXES AND TWO CORES, FIVE DISTRICTS CONNECTED IN ONE RING, CLEAR WATER BLENDING WITH GREEN PLANTS, AND ECOLOGICAL COEXISTENCE"

This design introduces Chaobai River landscape resources through the water network and the greenbelt to blend the two into the civic landscape and remodel the city pattern. The plan structure is as follows:

"Three axes" : three axes of urban development. The design relies on two landscape Avenues (Yingbin Avenue, Jiangtan Road) and one city axis Avenue (Xiaqi Road) to form three axes of urban development.

"Two cores" : two city cores. It is a Central Business Distinct (CBD) and a Recreational Business District (RBD).

"Five districts": five ecological residential districts. This is to form five residential districts around the city core, relying on the green axis and green zone.

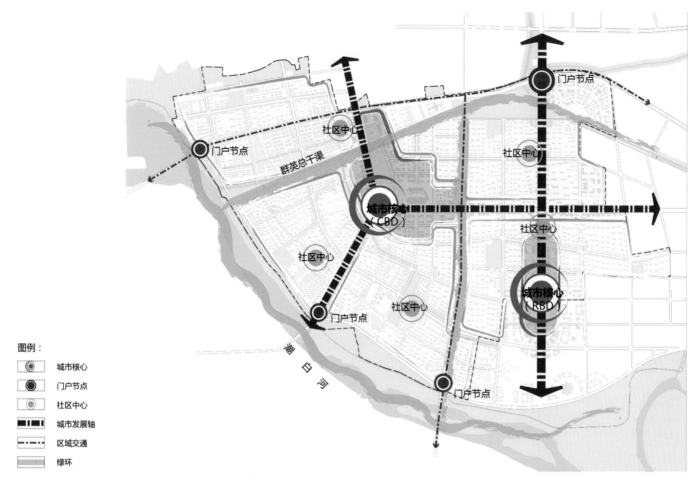

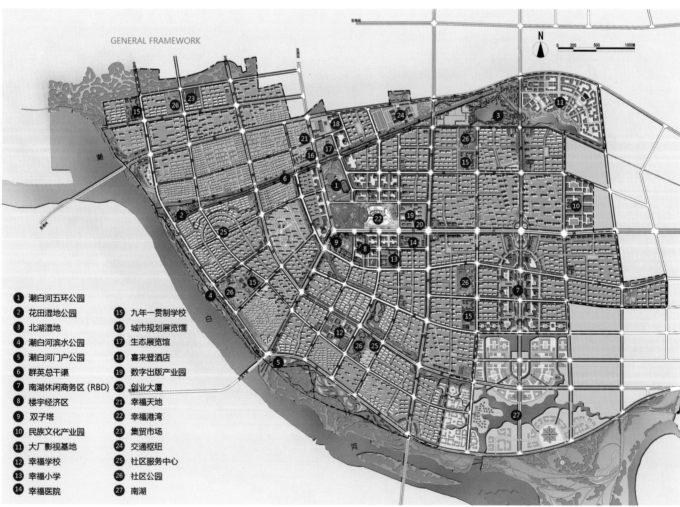

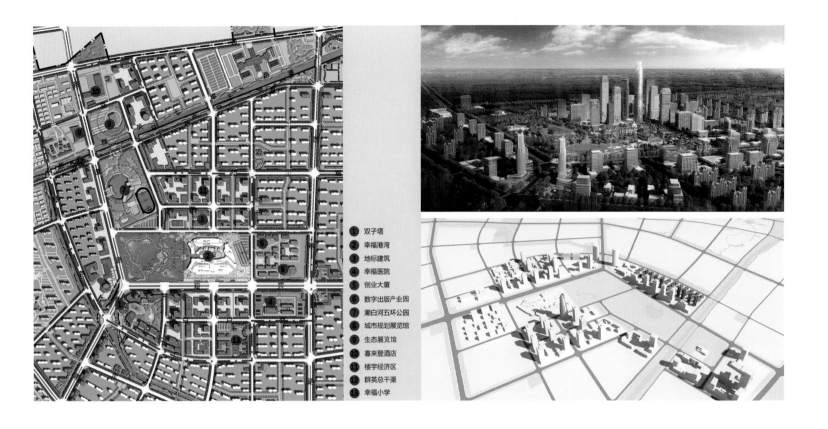

- ① 双子塔
- ② 幸福港湾
- ③ 地标建筑
- ④ 幸福医院
- ⑤ 创业大厦
- ⑥ 数字出版产业园
- ⑦ 潮白河五环公园
- ⑧ 城市规划展览馆
- ⑨ 生态展览馆
- ⑩ 喜来登酒店
- ⑪ 楼宇经济区
- ⑫ 群英总干渠
- ⑬ 幸福小学

039

THE PLANNING AND DESIGN OF SINO-GERMAN NEW INDUSTRIAL CITY IN TIEXI DISTRICT

铁西中德产业新城规划设计

Location
辽宁省沈阳市
Shenyang City, Liaoning Province

Area
31.7 平方千米
31.7 square kilometers

Time
2014 年至今
Since 2014

项目旨在打造"互联网+"时代与"中国制造 2025"时代背景下的中德创智天地、精工魅力核心。同时，它也是多元配套的智能制造产业服务中心、业态丰富的中德产业商务商业核心、魅力多姿的城市精英休闲娱乐中心。

"两核两轴三心五片"

规划提出"两核两轴三心五片"的格局：

在基地内部形成两大核心，分别为基地西部的产业研发核心和位于中央南大街及开发 25 号路的城市公共服务核心，产城融合，共促区域发展。

依托地铁 3 号线和 11 号线，形成两大城市发展轴，以开发 25 号路为主要城市发展轴，以中央南大街为次要发展轴。

创建多元化功能中心。在两大国际生活社区中，形成配套完善的社区服务中心及产业服务中心。

由西向东分别形成产业服务区、高端装备制造产业园区、国际生活社区、一处城市公共服务区和另一处国际生活社区。

The project aims to form a Sino-German innovation and intelligence district, a precision technology center in the contexts of "Internet Plus" and "China Manufacturing 2025". Meanwhile, it is also a multi-support service center for intelligence manufacturing industries, a business and commerce core for a variety of Sino-German industries, and an entertainment center for urban elite.

"TWO CORES, TWO AXES, THREE CENTERS, AND FIVE SECTIONS"

The planning brings up the layout of "two cores, two axes, three centers, and five sections":

This is to form two cores within the base. There are industry research core in the west and city public service core at South Central Avenue and No. 25 Kaifa Road, to achieve the integration of industry and city and promote regional development.

It will form two development axes based on No.3 and No.11 subway lines. No.25 Kaifa Road will be the primary city development axis, and South Central Avenue serves as the second development axis.

This is to build a diversified functional center. There will be a community service center and industry service center with complete facilities in the two international communities.

From west to east, it will form an industry service section, high-end equipment manufacturing industry park, one international living community, city public service section and the other international living community.

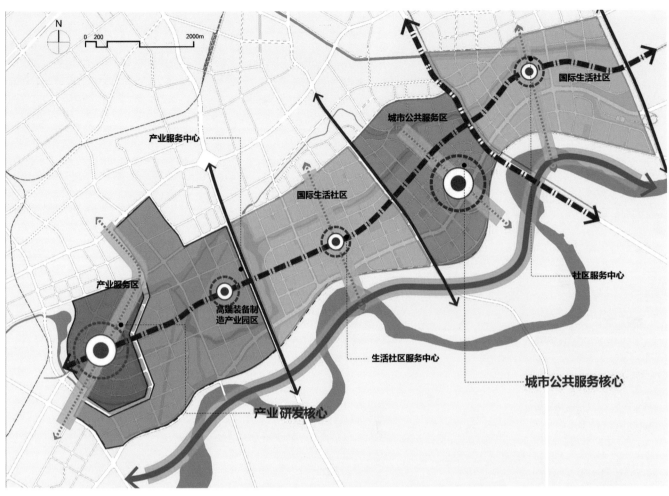

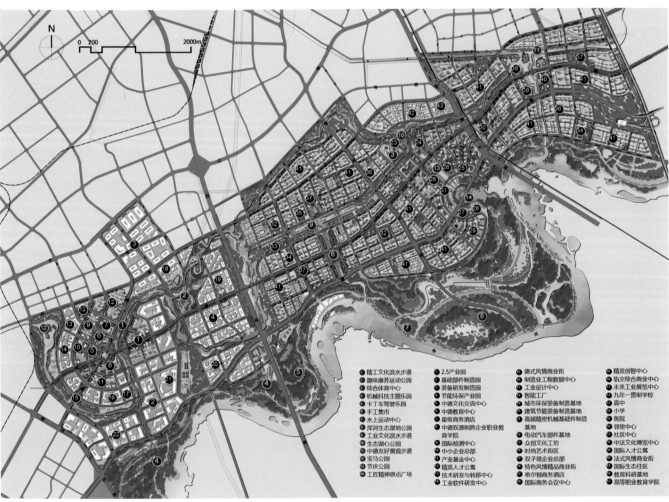

1.生态湖心公园
2.众创文化工坊
3.综合体育中心
4.时尚艺术街区
5.滨水城市休闲广场
6.下沉式集散广场
7.工业设计美术馆
8.工业文化纪念广场
9.商务休闲绿地广场
10.双子塔企业总部
11.市民休闲广场
12.节庆公园
13.特色风情精品商业街
14.工匠精神原点广场
15.城市创新中心
16.希尔顿商务酒店
17.国际商务会议中心
18.精英创智中心
19.轨交综合商业中心
20.城市综合交通枢纽站
21.P+R停车场
22.工业文明博物馆
23.未来工业展览中心

THE PLANNING AND DESIGN OF KANGYANG PARK TOWN IN DACHANG COUNTY

大厂县康养公园小镇规划设计

Location

河北省廊坊市
Langfang City, Hebei Province

Area

0.075 平方千米
0.075 square kilometers

Time

2016 年
2016

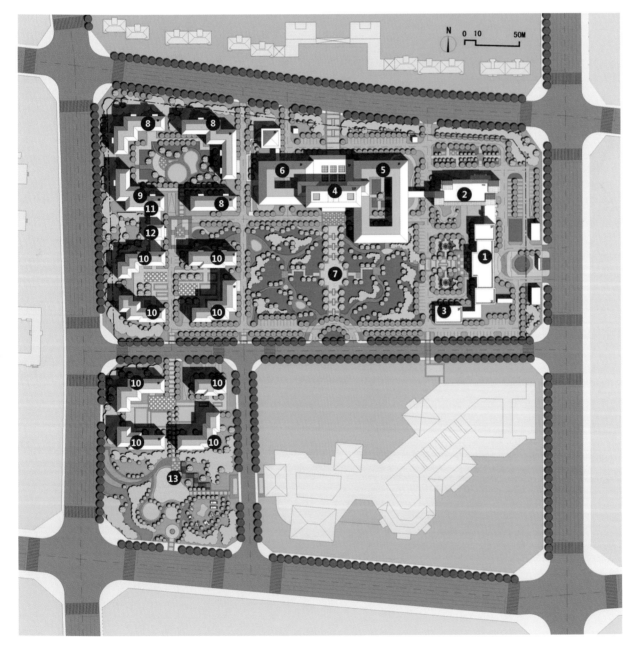

1. 门诊医技楼
2. 医技病房楼
3. 辅助设备楼
4. 医学康复中心
5. 健康促进中心
6. 生活护理院
7. 漫游百草园
8. 颐养公寓
9. 老年公寓
10. 老年住宅
11. 颐乐学园
12. 居家服务中心
13. 健身颐养公园

打造"医疗+"的创新产业模式,对接首都功能外扩

摆脱传统的医院建设模式,依托医疗设施建设,拓展相关健康产业的综合发展建设,打造创新的健康产业链。

缔造"医疗+康复+养生"的健康闭环项目新典范

北京以西地区山林生态较好,拥有多元的以生态为吸引力的康养项目,北京以东地区在这方面则相对不足。大厂被称为北方的水乡,生态资源良好,应该结合"以医促养"的理念,及时补充北京以东地区的不足。

创建本土化的养老及生活方式的新高地

随着我国老龄化程度的加深,全国各地都在建设养老项目,但很多项目都出现了"水土不服"的现象。我们立足于中国的文化特色,探索中国人自己的养老模式。

一环五片区,一轴多节点

一环:一个便捷功能环,串联各个功能板块,增强与基地南侧公园和商业的联系,形成一条便捷、功能丰富的步行流线。

五片区:即一个医疗片区、一个康复片区、两个养老社区和一个健身颐养公园片区。

一轴:结合中医知识,建设一条文化长廊景观轴线。

多节点:以多个功能节点和景观节点,服务整个片区。

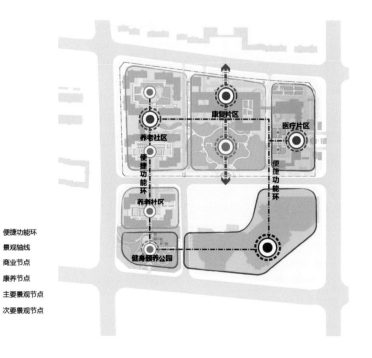

TO BUILD "MEDICAL TREATMENT PLUS" INNOVATION INDUSTRY MODE TO DOCK THE FUNCTION EXPANSION OF BEIJING

This is to expand comprehensive development in the related health industry to create an innovative health industry chain, based on the instruction of medical treatment facilities, by doing away with the traditional hospital construction.

CREATING A NEW MODEL OF CLOSED LOOP HEALTH PROJECT, "MEDICAL TREATMENT + REHABILITATION +REGIMEN"

Off the west of Beijing there is better mountain forest ecology than off the east and there have set up diversified regimen projects with eco environment as the attraction. Dachang County, renown as the watery place of the north part of China, has great ecological resources, and it should fill the gap with the concept of "promoting regimen by medical treatment".

CREATING A NEW HIGHLAND OF LOCALIZED ELDERLY CARING AND LIVING STYLE

With the increasing of China's elderly population, there is a wide spread in the building of caring centers for the elderly around the country, but many of which do not tally with the local situation. There is a need to explore a mode meeting the demands of the Chinese based on China's cultural characteristics.

ONE LOOP WITH FIVE AREAS, ONE AXIS WITH SEVERAL NODES

One loop means building a convenient function loop, connecting every functional section in series, extending the access to parks and business areas in the south of the base, to form a pedestrian line with convenience and rich functions.

Five areas include one medical treatment area, one rehabilitation area, two elderly caring areas and one exercise park.

One axis means to build a landscape axis of cultural corridor using traditional Chinese medicine knowledge.

Multi-nodes mean serving the whole area with several functional nodes and landscape nodes.

THE PLANNING AND DESIGN OF THE FEATURED TOWN —— TWO VALLEYS AND ONE VILLAGE

两谷一村特色小镇规划设计

Location	Area	Time
北京市张坊镇 Zhangfang Town, Beijing	约 0.35 平方千米 About 0.35 square kilometers	2016 年 2016

张坊镇是北京西南门户和北京郊区重要的商贸集散中心。京石第二高速公路开通后,这里与北京城区紧紧相连。穆家口村离北京市区只有 69 千米,距北京大兴国际机场 52 千米。

存古寨风采

穆家口村一共二百户村民都姓穆。穆家口村内有穆柯寨,至今仍然保留着昔日古寨的雄姿。

造美丽乡村

规划将穆家口村定位为体现"原生自然之美 • 乡土人文之魅"的京西风味十足的美丽乡村。

滨水休闲:依托生态特色,打造国际大都市中的世外桃源。

乡村体验:秉承乡村文化,诠释乡村民俗、乡村美食等元素,融合时尚与怀旧,打造乡村体验胜地。

乐活养心:倡导健康绿色的生活方式,聆听自己的心声,打造乐活养心地带。

山林娱乐:综合山林运动和娱乐,与穆柯寨联动,打造山林动感乐园。

由此形成三大功能分区,即滨水休闲区、原乡体验区和民宿风情区。

Zhangfang Town is the southwest gateway of Beijing and an important commerce and trade distribution center in the suburb of Beijing. After the opening of the second Beijing-Shijiazhuang Expressway, the town is closely connected to Beijing City. Mujiakou Village is only 69 kilometers away from Beijing and 52 kilometers away from Beijing Daxing International Airport.

SAVING THE STYLE OF THE OLD STOCKADE

A total of two hundred households are surnamed Mu in Mujiakou Village. There's a Muke Stockade, which still preserves the majesty of the old stockade.

BUILDING THE BEAUTIFUL VILLAGE

This project is to orientate Mujiakou Village as "a beautiful protogenetic village with local charm" embodying the characteristics of western Beijing.

Watery Leisure: Relying on ecological characteristics, it is going to be a fictitious land in the international metropolis.

Experiencing rural life: Adhering to the rural culture and combining rural folk and food, an integration of fashion and nostalgia is created. It is a resort for experiencing country life.

Happy Life and Healthy Mind: It is to advocate green and healthy lifestyle and following one's own heart, and create a zone to live a happy life and refresh the mind.

Entertainment in the Forest and Mountain: The forest sports and entertainment are integrated with Muke Stockade, so as to create a dynamic sports park in the forest and mountain.

In this way, three functional zones are created, i.e., the Watery Leisure Zone, the Original Rural Experiencing Zone and the Homestay Style Zone.

六大主题民宿

六大主题民宿【忘我·归隐山居】	吵子会（音乐主题）
	盖碗茶（茶艺主题）
	石头院里（石头主题）
	印年画（剪纸、年画主题）
	墨林山居（书画主题）
	酒腻子（古酒主题）

1. 吵子会
2. 盖碗茶
3. 石头院里
4. 印年画
5. 墨林山居
6. 酒腻子

THE PLANNING AND DESIGN OF THE FEATURED TOWN OF DAYU MOUNTAIN

大禹山特色小镇规划设计

Location

江苏省镇江市
Zhenjiang City, Jiangsu Province

Area

300.5 平方千米
300.5 square kilometers

Time

2015 年至今
Since 2015

基于大禹山现有的面貌与社会经济背景，紧扣当地的自然与产业特色，着眼未来进行规划。

"一核、四廊、四片区"

方案中的"一核"，指大禹山生态绿核。"四廊"，指四条生态绿廊。"四片区"，指四大组团依托大禹山生态空间和谐发展，形成一个活力组团、一个创智组团以及两个生态宜居组团。

The plan is to have a view for the future and meanwhile is based on the existing features and social economic background of Dayu Mountain, and draws on the characteristics of the local nature and industry.

"ONE CORE, FOUR CORRIDORS AND FOUR GROUPS"

"One core" in the plan is the core of Dayu Mountain ecological greenland. "Four corridors" refer to the four ecological green corridors. "Four groups" mean the four large groups based on Dayu Mountain ecological space. The four large groups develop harmoniously, and work together to form a vitality group, an innovation group and two ecological residence groups.

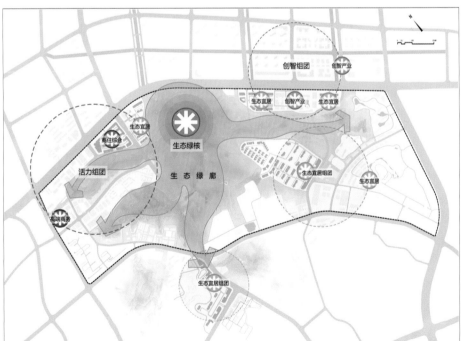

THE PLANNING AND DESIGN OF THE NEW INDUSTRIAL TOWN IN DACHANG COUNTY

大厂县产业新城规划设计

Location

河北省廊坊市
Langfang City, Hebei Province

Area

4.7 平方千米
4.7 square kilometers

Time

2014 年至今
Since 2014

水润如意地，绿满大厂城

"水润"聚焦水环境，意为雨水滋润万物；"如意"象征着美好、幸福，寓意为如意相随。以"如意"为设计主题，以水、绿为线索，将"如意"的内涵渗透到整个规划中。

"绿满"点明了绿色健康的环境，由公园、水、社区绿地等，构成点、线、面立体式的绿化环网。绿网相连，蓝脉相绕，点轴渗透，创建出"虽由人造，宛若天开"的一座新城。

"一轴、一带、六心"，缔造城水交融、多元一体的城市新貌

"一轴"：由民族宫、中央公园、行政中心、商务中心、博物馆、科技馆等串联成新城的核心发展轴，也是景观核心轴。

"一带"：沿鲍丘河打造的重要的生态景观带。

"六心"：集商业、贸易、文化、休闲、商务等多功能于一体的六大功能核心。

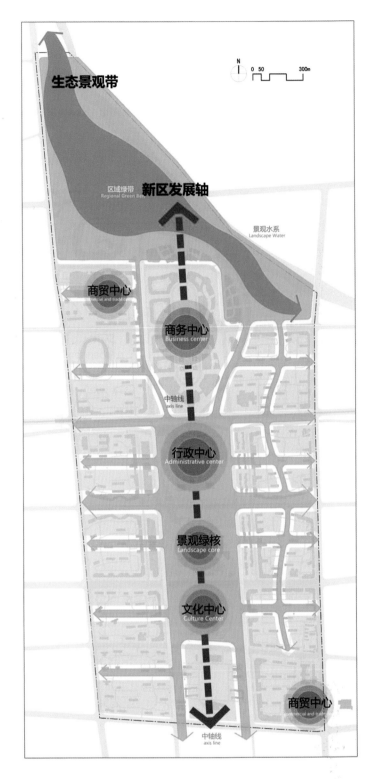

WATERY NOURISHMENT OF CONTENTED (RUYI) PLACE, THE GREEN FLOURISHES IN DACHANG COUNTY

The "watery nourishment" is to focus on the watery environment, signifying that the rain is coming down to nourish the nature; "Ruyi" symbolizes being surrounded by the beautiful and happy things. Taking "Ruyi" as the theme, the project uses water and green as clues to embody the connotation of contentment into the entire design.

"The green flourishes" points to its green and healthy environment, with the park, water, and community green spaces contributing to point, line and plane three-dimensional green networks, circling the whole town. The green networks join together, the water surrounds and connects the town, and the points and axis are combined, building a new town which is "arterially created but very natural."

"ONE AXIS, ONE BELT AND SIX CORES" ARE USED TO CREATE NEW LANDSCAPE OF WATERY CITY WITH COMPREHENSIVE FUNCTIONS

"One axis": the National Palace, central park, administrative center, business center, museum, science and technology museum are connected in series to form the core development axis of the new town, and it becomes the core axis of the landscape as well.

"One belt": an important ecological landscape belt along the Baoqiu River.

"Six cores": they are the six significant functional cores, integrating the multitudinous functions of business, trade, culture, leisure and commerce, etc.

筑核 CORE →	透绿 GREEN →	营城 CITY
"核心引领，区域引擎"	"融合渗透，绿色生态"	"有机聚合，活力新区"

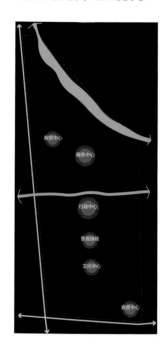 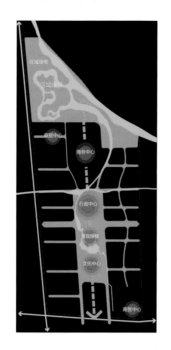 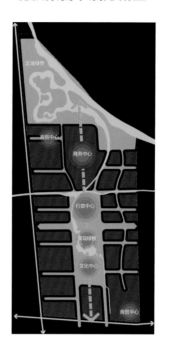

水润如意地
绿满大厂城

THE PLANNING AND DESIGN OF THE NEW MANRONG INDUSTRIAL TOWN

Location
辽宁省沈阳市
Shenyang City, Liaoning Province

Area
约 2.27 平方千米
About 2.27 square kilometers

Time
2014 年至今
Since 2014

迎合生态需求，对接区域一体。产业园区互动，营造特色魅力。规划的目标在于打造最具吸引力的活力新城。

"一轴、一带、三组团"

方案形成了如下布局：

"一轴"：沿桂花街打造的发展轴是新城形象重要的展示界面。

"一带"：沿浑河形成集休闲、娱乐、观光等功能于一体的魅力水岸带。

"三组团"：文创产业组团、文教水岸社区和国际水岸社区。

The plan meets the ecological needs and links up with the regional integration. It will create a unique charm by making industrial parks interact with each other. The plan aims to create the most attractive new town with vitality.

"ONE AXIS, ONE BELT AND THREE GROUPS"

The design has formulated the following layout:

"One axis" is the Development Axis, which is built along the Guihua Street and becomes an important display interface of the new town's image.

"One belt" is the Leisure Belt which combines the functions of leisure, entertainment and sightseeing along the Hunhe River.

"Three groups" include Cultural and Creative Industries Group, Educational Waterfront Community and International Waterfront Community.

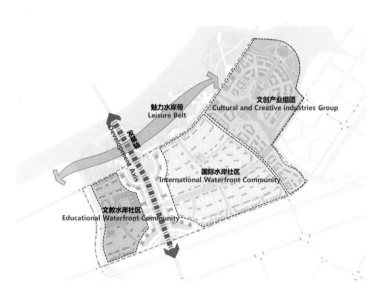

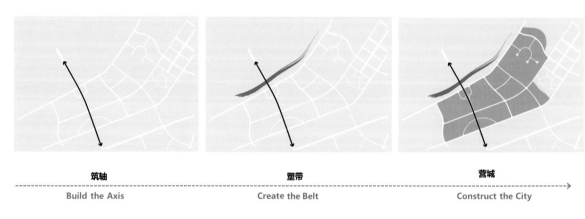

筑轴	塑带	营城
Build the Axis	Create the Belt	Construct the City
沿桂花街打造发展轴，布置城市公共功能，树立城市新形象。	充分利用现有自然资源优势，沿浑河塑造一条国际化魅力水岸带。	以发展轴、魅力水岸带及三环线为骨架，在三环高速北侧构筑城市空间。

1. 文化休闲艺术园
2. 湿地雨水公园
3. 儿童游乐园
4. 都市沙滩
5. 文化传媒商务港
6. 新媒体数字科技港
7. 滨水休闲商业街
8. 幸福港湾风情商业街
9. 生产力促进产业港
10. 创意天地文化传媒港
11. 彩虹桥
12. 文化创智工坊
13. 国际文化交流中心
14. 生态宜居社区
15. 幼儿园
16. 教堂
17. 小镇中心（五合一会所）
18. 国际学校

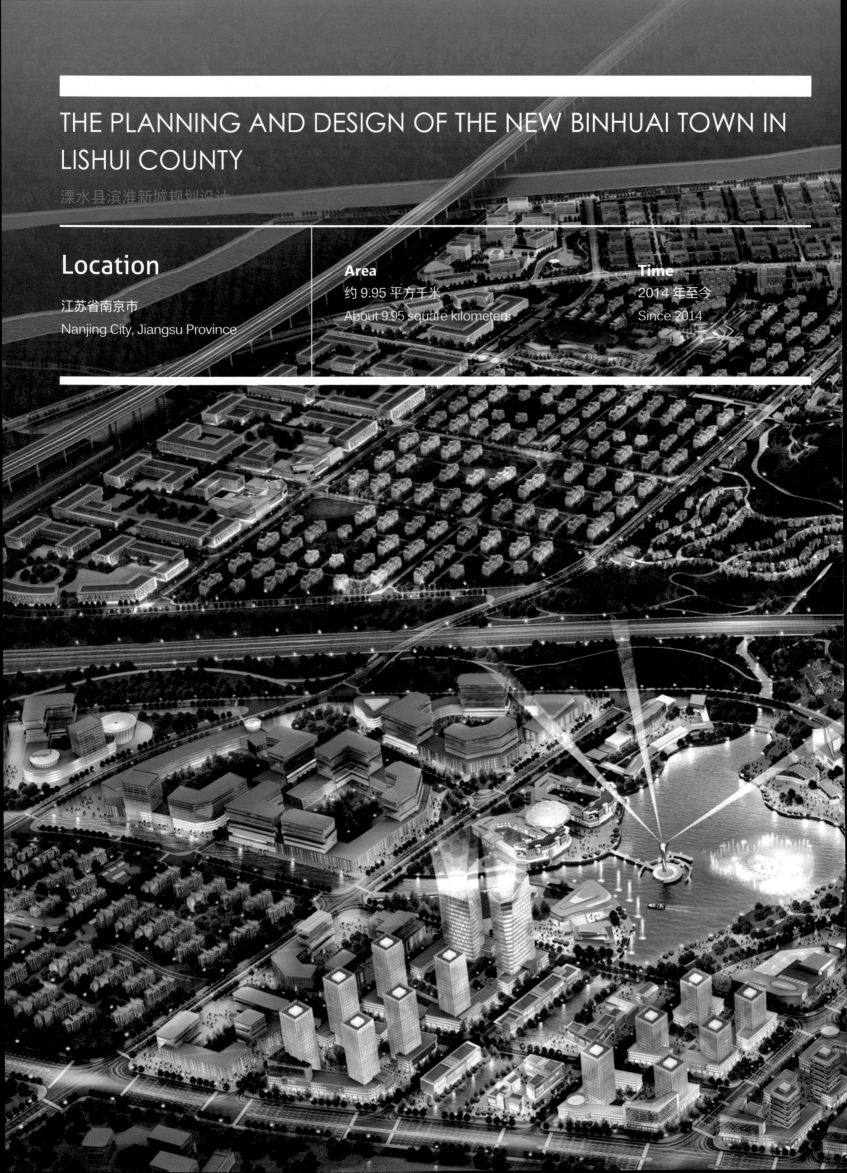

THE PLANNING AND DESIGN OF THE NEW BINHUAI TOWN IN LISHUI COUNTY

溧水县滨淮新城规划设计

Location
江苏省南京市
Nanjing City, Jiangsu Province

Area
约 9.95 平方千米
About 9.95 square kilometers

Time
2014 年至今
Since 2014

凭借自身的综合优势，位于江苏省南京市溧水县的滨淮新城正蓄势待发。

"一核、两翼、双心、多廊道"

规划方案提出，以生态核心为引导，以绿网串联，形成"一核、两翼、双心、多廊道"的空间结构。

"一核"：基地中部赵山、团山相拥而立，构筑新城生态核心。

"两翼"：宁高高速两侧城市功能区，东西联动，全面对接外围产业及城市功能。

"双心"：东侧科创中心与西侧服务中心双核驱动，强力支持区域持续高速发展。

"多廊道"：以水绿融城构建新城生态基础，结合山水环境资源，构筑生态廊道。

功能与组团

根据规划，十个组团依照服务、生活、科研、产业的顺序向外分布。这十个组团又分为四个类别，分别为绿色健康（休闲体育组团）、科创研发（创智组团1、创智组团2）、产业创新（智能制造组团1、智能制造组团2、智能制造组团3）和城市生活（综合功能组团、城市宜居组团、城市水系组团1、城市水系组团2）。

"ONE CORE, TWO WINGS, TWO CENTERS AND MULTIPLE CORRIDORS"

The plan proposes that the development should be guided by an ecological center and connected by green networks, to form a spatial structure of "one core, two wings, two centers and multiple corridors."

"One core": there are two mountains, Mount Zhao and Mount Tuan, in the central base, which can be used to build the ecological core of this new town.

"Two wings": it refers to the urban functional areas located on both sides of Ninggao Expressway, which form a linkage of the west and the east, and connect comprehensively the outside industries and urban functions.

"Two centers": it means dual-core drive of the eastern science and innovation center and the western service center, to provide a strong support for this rapid regional sustainable development.

"Multiple corridors": it refers to establishing ecological base of the new town with water and trees, and by combining the landscape resources, to form ecological corridors.

FUNCTIONS AND GROUPS

According to the plan, the ten groups are distributed from inside to outside in the order of service, life, scientific research and industry. The ten groups are classified into four types, including green and healthy (Recreational Sports Group), science, innovation research and development (Innovation Group 1, Innovation Group 2), industrial creation (Intelligent Manufacturing Group 1, Intelligent Manufacturing Group 2, Intelligent Manufacturing Group 3) and urban life (Comprehensive Functional Group, Urban Livability Group, Urban Watery Group 1, Urban Watery Group 2).

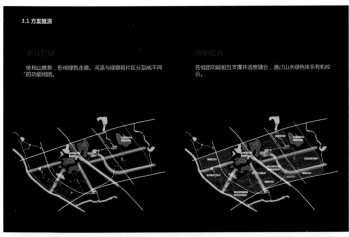

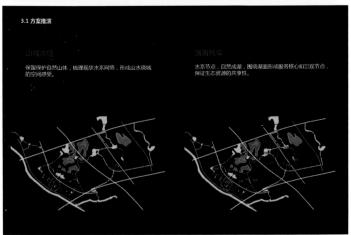

山体景观：团山、角山
廊道：山与山、山与水之间的绿色通廊，道路景观通廊
界面：沿山、沿水、沿高速公路的建筑及景观界面
节点：由水、绿树和广场组成的各个景观节点
地标：面朝新木塘公园，树立整个地块朝气蓬勃的建筑形象

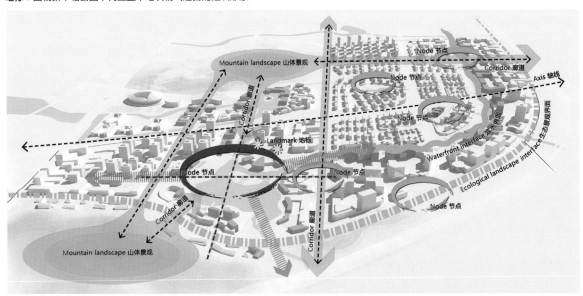

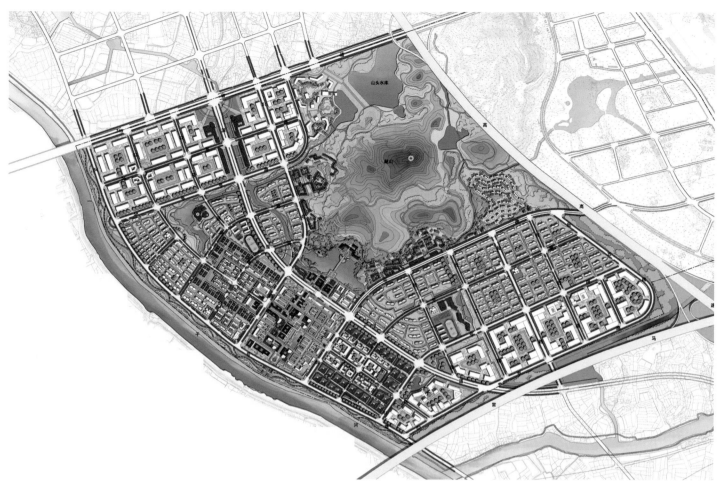

THE PLANNING AND DESIGN OF THE NEW INDUSTRIAL TOWN IN HUAILAI COUNTY

怀来县产业新城规划设计

Location
河北省张家口市
Zhangjiakou City, Hebei Province

Area
约 11.51 平方千米
About 11.51 square Kilometers

Time
2014 年至今
Since 2014

怀来县地处河北省西北部，东邻北京，西接张晋蒙。规划旨在打造北京未来的养生之地，创建一个山水宜居的怀来。总体上，以"山、水"为核心资源，以"葡萄园景观"为个性特征，发展生态产业，让山城相映，城湖共生。

规划结构：发展新城公共服务核心，创造城市高端生活区，以生态廊道联系山水

规划区位于怀来新兴产业示范区东北部。规划区主要是城市功能区的生态居住组团和城市核心组团。城市公共服务中心位于基地南部，城市发展轴与生态绿廊穿基地而过。基地主要为居住、商业等城市功能用地。

"一轴、两核、九片区"

"一轴"：城市发展轴，南北向串联城市商业商务服务核心、老年颐养服务核心和各大片区，是城市形象的重要展示面。

"两核"：包括城市商业商务服务核心，以及老年颐养服务核心。其中，城市商业商务服务核心位于基地南部地块，是基地的城市核心，提供前沿的娱乐、购物、商务、生活服务体验，是新城的重要功能核；老年颐养服务核心位于基地北部地块，是养老、养生的活力服务核，提供健康、颐养、社交、居住服务等体验。

"九片区"：规划区内九大不同主题特色的居住片区，主要包括：亲水宜老主题居住片区、乐活养老主题居住片区、智乐养老主题居住片区、运动康老主题居住片区、绿色康养主题居住片区、葡萄庄园主题居住片区、生态宜居主题居住片区、山前休闲主题居住片区和居民安置居住片区。

Located in the northwest of Heibei Province, Huailai County borders Beijing in the east and to the west is Zhangjiakou City, Shanxi Province and Inner Mongolia Autonomous Region. This planning aims to develop a future resort for Beijing citizens and create livable Huailai with mountains and watery systems. The general plan is to feature the "mountains and waters" as the core resource, to take the "vineyard" as the characteristic landscape, to establish eco-agriculture, with the city and the mountains highlighting each other, and the city and the lake coexisting in harmony.

THE PLANNED STRUCTURE: DEVELOPING THE PUBLIC SERVICE CORE OF NEW HUAILAI, CREATING A HIGH-END URBAN RESIDENTIAL AREA, USING AN ECO CORRIDOR TO CONNECT THE MOUNTAINS AND WATERY SYSTEMS

The planned area is located in the northeast of new Huailai industrial demonstration area. The planned area mainly includes the eco residential groups of urban functions and the core urban groups. Urban public service center is located in the south of the base, with the city development axis and eco green corridors passing through. The "base" is mainly for the land use of residential, commercial and other urban functions.

"ONE AXIS, TWO CORES, NINE SECTIONS"

"One axis": the city developing axis. It passes through urban business and commerce service center, the elderly resident service center and all the other areas in south and north direction, highlighting the best image of Huailai to the world.

"Two cores": the business and commerce service center and the elderly resident service center. The former is located in the south of the base, and functions as the urban center of the base, providing forefront entertainment, shopping, commerce, and life service experiences. It is an important functional center of the new city. The latter is located in the north of the base, and works as the base's vitality center for pension and health, providing health, enjoyment, social life, residence services, etc.

"Nine sections": residence section with nine different themes and features in the planned area: water intimacy and friendly to the elderly, happy elderly, intellectual elderly, sports elderly, green health, vineyard theme, eco livability area, mountain foot recreational theme, and relocation section.

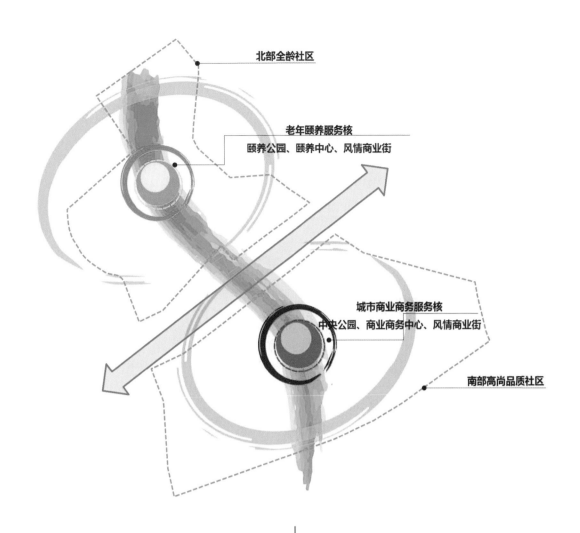

- 北部全龄社区
- 老年颐养服务核　颐养公园、颐养中心、风情商业街
- 城市商业商务服务核　中央公园、商业商务中心、风情商业街
- 南部高尚品质社区

一轴 · 两核 · 九片区的空间模型的建立

以高速公路为界，划分南北两大板块

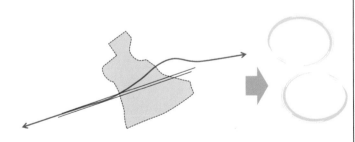

以河道冲沟为轴，建立绿色联系廊道

以板块服务为核，建立活力板块中心

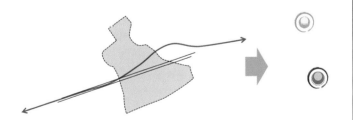

以用地规模为度，划分九个主题居住片区

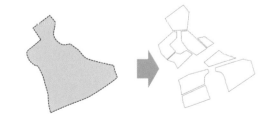

1. 颐养公园
2. 中央公园
3. 颐养中心
4. 风情商业街
5. 西榆林新民居
6. 商业综合体
7. 商业商务中心
8. 风情小镇
9. 瑞云酒堡
10. 行政中心
11. 医院
12. 中学
13. 小学
14. 亲水宜老主题 居住片区
15. 乐活养老主题 居住片区
16. 智乐养老主题 居住片区
17. 运动康养主题 居住片区
18. 绿色康养主题 居住片区
19. 葡萄庄园主题 居住片区
20. 生态宜居主题 居住片区
21. 山前休闲主题 居住片区
22. 居民安置 居住片区
23. 文化展览中心
24. 酒店
25. 公交巴士总站
26. 幸福学校
27. 体育公园

THE PLANNING AND DESIGN OF THE FILM AND TV TOWN IN DACHANG COUNTY

大厂县影视小镇规划设计

Location

河北省廊坊市
Langfang City, Hebei Province

Area

0.63 平方千米
0.63 square kilometers

Time

2014 年
2014

大厂县位于北京发展主轴的收头位置，北京东部发展轴与首都经济圈的十字交汇处，拥有不在北京、胜在北京的地理区位。在京津冀一体化发展的时代背景下，大厂县未来的发展有许多机会与可能。

为全面发展影视高新技术产业推波助澜

项目的定位在于打造全球影视高新技术实践区。产业定位是涵盖电视节目的创意策划、前期拍摄、后期制作；涵盖数字化体验、教育培训等领域并辐射衍生品制造、旅游体验等产业环节。

规划的构思在于打造多元尺度空间，实现创意和创新的集合。每个区均植入不同尺度的内部空间，如适合交流的、适合创作的、适合展演的、适合拍摄的，形成活力酷炫的魅力场所。

一个中心、一个广场加多个组团

在格局方面，呈现为一个中心、一个广场加多个组团的有机关系，即演艺中心、星空广场，以及CCTV组团、世纪汉唐组团、办公组团、孵化器组团、BASE动效研发组团、展示中心组团。

Dachang County is located at the end of the Beijing development axis and the intersection of the east Beijing development axis and capital economic zone. It has a geographical location to the effect of "being not in Beijing but being better than in Beijing". Against the background of integrated development of Beijing-Tianjin-Hebei Region, Dachang County has more opportunities and possibilities in store.

PROMOTING THE DEVELOPMENT OF FILM AND TELEVISION HIGH-TECH INDUSTRY

The program is to build a high-tech practice area for global film and television industry. The industry orientation includes creative planning, pre-shooting, post-production of television programs; it also covers areas such as digital experience, education and training, and industrial sectors such as derivatives manufacturing, tourism experience and so on.

The planning is to create a multi-dimensional space, to achieve a convergence of creativity and innovation. Each district will be embedded with the interior space with different scales to form dynamic and vigorous places, such as spaces suitable for communication, for creation, for the show, or for film shooting.

ONE CENTER, ONE SQUARE AND MULTIPLE GROUPS

The layout is to have one center, one square and multiple groups integrated, namely the performance center, the star sky square, as well as CCTV group, the Han and Tang dynasties group, office group, incubator group, BASE animation effect research and development group, and the exhibition center group.

区域发展背景 REGIONAL DEVELOPMENT BACKGROUND

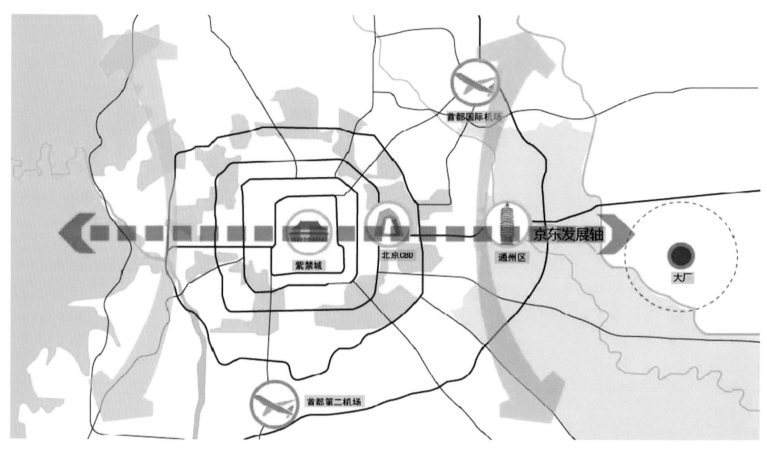

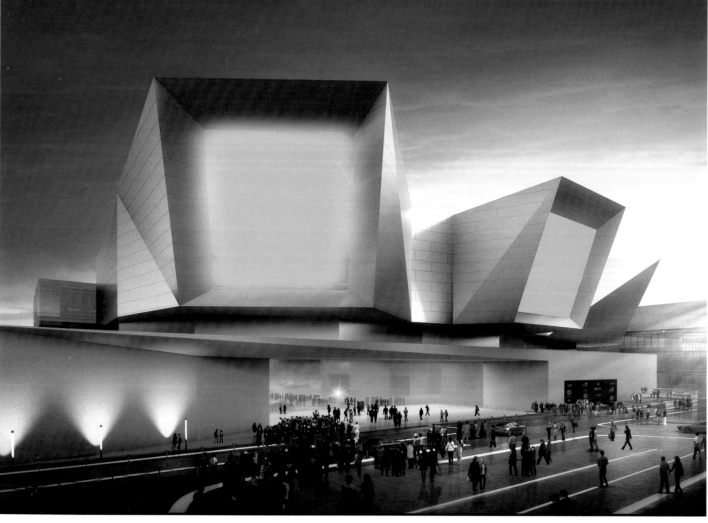

THE PLANNING AND DESIGN OF THE EMERGING INDUSTRY NEW TOWN IN GU'AN COUNTY

固安县新兴产业新城规划设计

Location	Area	Time
河北省廊坊市 Langfang City, Hebei Province	18.96 平方千米 18.96 square kilometers	2014 年至今 Since 2014

现代都市中的人们都在梦想着这样的生活：远离喧嚣、私密自在、品质上乘、丰富多维、安全健康……本规划的核心目标是将固安县新兴产业示范区的城市功能区打造成为实现人们这一梦想的魅力之地。

建立核心，打造绿网，营造特色社区

本规划提出，建立多个特色发展核心，依托主要规划道路打造绿网系统，打通永定河和引清干渠两大生态资源与规划基地的连接，以绿网串联，形成魅力空间和特色公共活动体系。通过创建最强城市核心、发展生态绿网、打造风情街道与特色社区记忆点，打造最具场所感的公共领域。

Modern city dwellers dream of such a life: being away from the hustle and bustle, to have privacy, freedom, high quality, colorful activities as well as security and health... The core objective of this plan is to make city functions of the new Gu'an industrial demonstration area into a charming place that realizes people's dream.

ESTABLISH CORE AREA, BUILD GREEN NETWORK, AND CREATE A COMMUNITY WITH SPECIAL FEATURES

The plan proposes to establish a number of featured development core areas, build green network system along the planned main road, connect the two main ecological resources of Yongding River and Yinqing trunk canal with the planned base, and to form a charming space and a special public activity system by the green network connecting in series. Through establishing the strongest city core area, developing ecological green network, it will create stylistic streets and featured community memory points, to achieve the creation of a public domain with the most powerful sense of place.

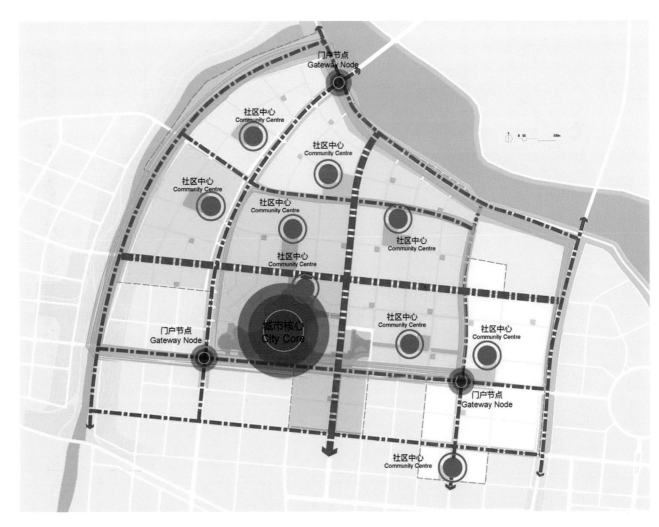

093

THE PLANNING AND DESIGN OF THE NEW INDUSTRIAL TOWN IN WEN'AN COUNTY

文安县产业新城规划设计

Location
河北省廊坊市
Langfang City, Hebei Province

Area
24.4 平方千米
24.4 square kilometers

Time
2014 年至今
Since 2014

规划目标:打造区域经济支点、文安发展核心、生态产业引擎

未来,文安将不断做大做强主导产业,并以此为基础形成制造业核心产业;以发展第三产业为途径,培养新城大产业链经济;以发展和培养高端制造业和新型服务业为重点,形成高端产业经济,并最终形成"产业金字塔"。

在主导产业的带动下,文安上下游和相关产业将迅速发展。企业的增加带动人口的增加,并带动城市整体功能、配套设施的完善和升级。城市的升级又将再次催生产业的升级,形成"以业兴城,以城促业"的发展格局。

规划格局:两轴一带,三心并举

"两轴":一是迎宾大道发展轴,它也是产业新城的城市形象展示轴;二是文左路产业轴,是连接文安老城的产业发展轴。

"一带":生态文化带,串联产业区、生活区及商业区的重要景观带。

"三心":一是城市核心,指集商业、商务、文化、休闲、旅游、生活于一体的城市核心区;二是生态绿心,指为周围生活组团服务的城市文化生活核心,围绕景观大湖形成的大型生态绿地;三是产业核心,它是产业形象展示区,包含企业总部、小企业孵化园、创新研发园等诸多功能构成部分。

PLANNING OBJECTIVES: TO BUILD REGIONAL ECONOMIC FULCRUM AND DEVELOP WEN'AN COUNTY CORE WITH ECO INDUSTRY AS THE ENGINE

From now on, Wen'an is to continue to forge bigger and stronger leading industry, and formulate the manufacturing core industries based on it. Taking the tertiary industry as a development approach, the county aims to cultivate large industrial chain economy in the new town. Giving priority to develop and cultivate high-end manufacturing and new services, it is to form high-end industrial economy and ultimately establish the "industrial pyramid".

Driven by the leading industry, Wen'an will achieve rapid development in related industries upstream and downstream. The increase in the number of enterprises promotes the increase of population, the improvement and upgrading of the general functions and the supporting facilities of the city. The upgrading of the city will spawn the upgrading of the industry in turn, forming the development pattern of "industry brings prosperity to the city, and the city promotes the industry".

PLANNED PATTERN: TWO AXES AND ONE AREA, THREE CORES DEVELOP SIMULTANEOUSLY

"Two axes": one is the development axis at Yingbin Avenue, which is also the city image display axis in the new industrial town; the other is the industrial axis on Wenzuo Road, which is the industrial development axis connected to the old town in Wen'an County.

"One area": the ecological culture area, connecting the important landscape belts in tandem such as industrial parks, residential and business districts.

"Three cores": one is the city core, which refers to the urban core areas integrating commercial, business, culture, leisure, tourism and residence; the second is the ecological green core, referring to the urban cultural life core serving for the surrounding residential group which forms a large ecological green space around the landscape of the Great Lakes; the third is the industry core, which is the industrial image display area, including many functional components such as corporate headquarters, small business incubator, innovation and development park.

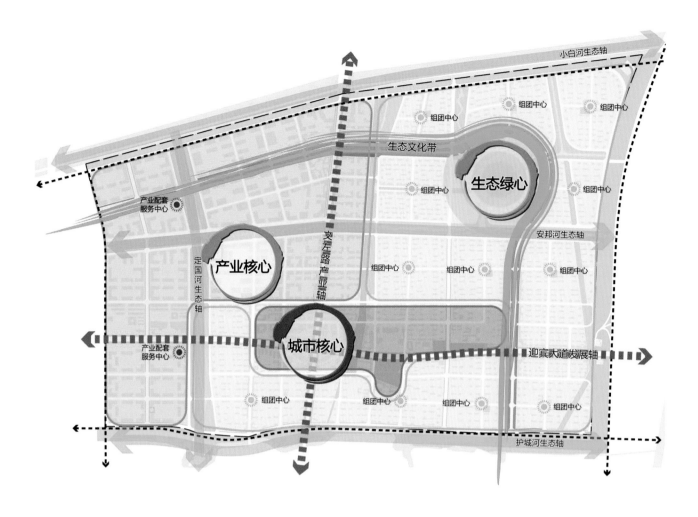

THE PLANNING AND DESIGN OF THE NEW INDUSTRIAL TOWN IN THE NORTH OF THE GRAND CANAL

大运河北产业新城规划设计

Location
河北省廊坊市
Langfang City, Hebei Province

Area
6.13 平方千米
6.13 square kilometers

Time
2014 年至今
Since 2014

规划传承大运河的历史文化，承接北京"滨水而居的城市"理想生活，形成北京首个具有水韵的体验式综合文化休闲水乡。方案以"滨水宜居、文化嫁接、水韵商业、体验互动"为关键词。

"一环双水岸，一轴两片区"

东西一条多彩水环，集商业娱乐、文化休闲和生态居住于一体，同时形成区域商业区核心。

东西两大水岸，提升基地价值。东岸运河文化水岸公园，西岸生态湿地水岸公园，各有特色。一条城市中央景观轴将劣势变为优势，形成贯通南北的景观通廊。南北两大片区功能完整，组团清晰，各配两个社区级服务中心。

The plan is to inherit the history and culture of the Grand Canal, undertake the Beijing ideal life of "waterfront city", and constitute the first comprehensive cultural leisure watery town with water rhyme experience in the Beijing area. The key words of the program are "livability waterfront, cultural integration, water rhyme business, interaction experience".

"A BELT OF TWO WATERFRONTS, ONE AXIS WITH TWO REGIONS"

A colorful water belt integrates business entertainment, culture and leisure, and ecological living. And along with it, a regional business district core is formed.

Two major waterfronts of east and west coasts upgrade the base value. The east canal culture waterfront park and the west eco-wetland park are designed with distinct characteristics. A city central landscape axis turns its disadvantages into advantages, forming a landscape corridor running through the north and the south, Both northern and southern regions have complete functions, and have their distinctive means of group formations, each equipped with two community-level service centers.

THE PLANNING AND DESIGN OF THE CORE DISTRICT OF NEW INDUSTRIAL TOWN IN JIASHAN COUNTY

嘉善县产业新城核心区规划设计

Location	Area	Time
浙江省嘉兴市 Jiaxing City, Zhejiang Province	约 0.1 平方千米 About 0.1 square kilometers	2014 年至今 Since 2014

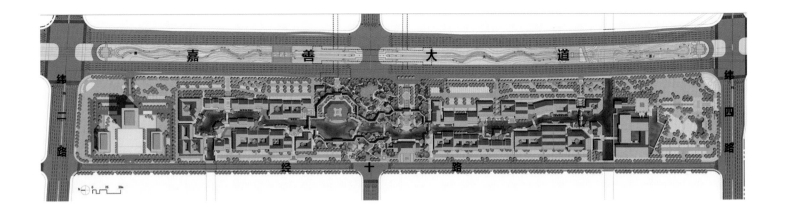

嘉善地处长三角城市群核心区域,是浙江省接轨上海的第一站,素以鱼米之乡、丝绸之府、文化之邦而闻名。

打造水乡特色的新城核心区

规划立足于江南水乡的地理文化和嘉善的发展现状,打造极富商业活力与文化活力的新城核心区。

横向规划:一条水面、两条水街。

纵向规划:一核、两翼、三入口。

节点规划:一大三小四个演出场、一横一转五纵七个对景轴。

Jiashan, located in the core region of the Yangtze River Delta city cluster, is the first stop for Zhejiang to connect Shanghai, and has been known for its rice and fish, silk, and culture.

BUILDING A NEW CORE ZONE WITH WATERY TOWN FEATURES

The plan is based on the geographical culture of the watery town and the development conditions of Jiashan, to build a core district of the new town that shows both commercial and cultural vitality.

Horizontal planning: a river and two commercial streets alongside the river.

Longitudinal planning: one core, two wings and three entrances.

Node planning: four performance venues: a large one and three small ones; seven landscape axes: one horizontal, one curved and five longitudinal.

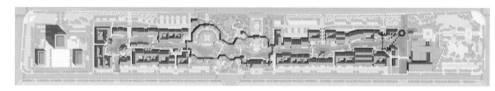

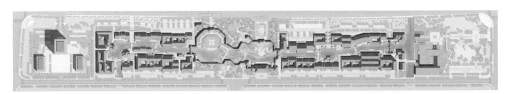

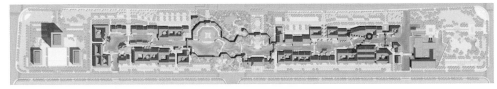

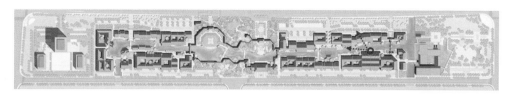
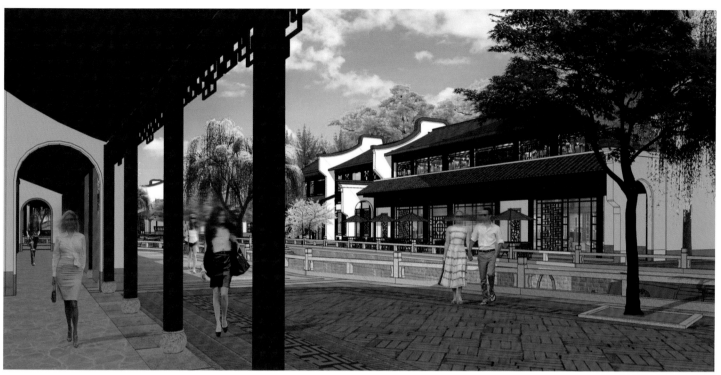
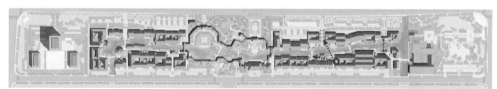

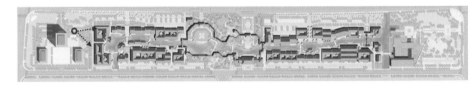

THE PLANNING AND DESIGN OF THE PEACOCK CITY IN WESTERN GU'AN COUNTY

固安县西部孔雀城规划设计

Location	Area	Time
河北省廊坊市 Langfang City, Hebei Province	约 0.94 平方千米 About 0.94 square kilometers	2014 年至今 Since 2014

在美丽的固安，一座"孔雀城"正等待绽放其独特的魅力。

规划理念：能效与魅力并存

规划旨在打造最佳精英生活区，塑造极具魅力的风情小城，一个优雅而又不失能效的精英之城。

通过多元魅力元素的注入，持续提高基地开发价值，营造出多维、可观、可触的城市魅力体系。

In beautiful Gu'an, the "peacock city" is waiting to show its unique charm.

PLANNING CONCEPT: A COMBINATION OF EFFICIENCY AND CHARM

This plan aims to create the best living area for the elites and to build a city with cultural attractions, a place with elegant and efficient functions.

By ejecting multi-charm elements, the design is to continuously improve the development value of the base and create a charming city system that is multi-dimensional, visible and tangible.

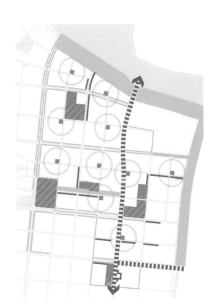

一条回家路
家的归属感

三条动感绿楔
滨水风光、折叠景观与都市绿丘的动感

四个社区中心
服务与绿色的共享

五个主题公园
静享都市生活，品味自然之美

十个邻里交流空间
最近的距离，最宜人的场所

多元邻里商街
有的艺术时尚，有的散发书卷气息，各具特色

绿色运动网络
车行的感受，步行的享受，绿色与运动的叠合

THE PLANNING AND DESIGN OF THE BEAUTIFUL VILLAGE OF LIRANGDIAN IN GU'AN COUNTY

固安县礼让店乡美丽乡村规划设计

Location	Area	Time
河北省廊坊市 Langfang City, Hebei Province	约 0.61 平方千米 About 0.61 square kilometers	2015 年至今 Since 2015

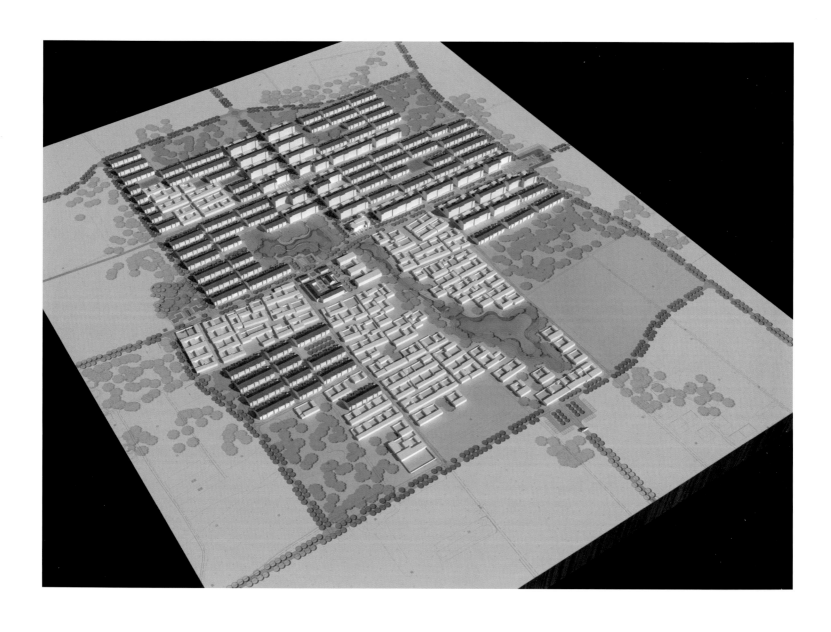

规划致力于将固安县打造成为全国领先的现代农业产业化示范基地、河北省绿色健康的美丽乡村生活新典范，以及京南最具地方特色的乡村体验地。

"一轴、两心、一带、两片区"，精准布局

"一轴"：确立礼让店支线的乡镇发展轴。

"两心"：确立以乡政府、学校为核心的综合服务核心；新建以滨水广场、商业为核心的生活休闲核心。

"一带"：建设滨水休闲景观地带。

"两片区"：分步骤建设南北两个生活片区。

在启动区，形成"一核、一轴、四片区"的方案。

"一核"：在中心布置幼儿园、便民服务中心、卫生室、农资超市等，形成公共服务核心。

"一轴"：新农村景观形象轴，它从村庄正中穿过，展现农村风貌，同时向西联系沙河口村。

"四片区"：四大居住片区。

同时，还将保留现状林地，通过相互连接形成中央公园，切实提高居民生活品质。

The plan is committed to turning Gu'an County into a modern agriculture industrialization demonstration base taking a national leading position, a new model of beautiful village life with green and health in Hebei Province, and a place to experience the most distinctive local rural features to the south of Beijing.

"ONE AXIS, TWO CORES, ONE ZONE & TWO DISTRICTS", AN ACCURATE LAYOUT

"One axis": to establish the township development axis of Lirangdian Village branch.

"Two cores": to establish an integrated service core of the village government and school; to establish a new leisure life core of the waterfront square and business.

"One zone": to construct a zone of waterfront leisure landscape.

"Two districts": to construct two residence districts in the north and south in steps.

The starting area is to execute the design of "one core, one axis and four districts".

"One core" : to set up kindergartens, convenience service centers, health clinics, agricultural supermarkets, etc., forming the public service core in the center of the village.

"One axis" is the image axis of new rural landscape, going through the middle of the villages, demonstrating the rural features while connecting Shahekou Village to the west.

"Four Districts" is the four big residence districts.

Meanwhile, it will preserve the forest lands and connect them to form the Central Park, so as to effectively improve the quality of the life of the residents.

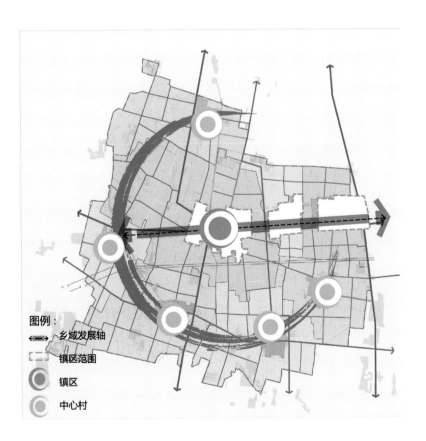

图例：
- 乡域发展轴
- 镇区范围
- 镇区
- 中心村

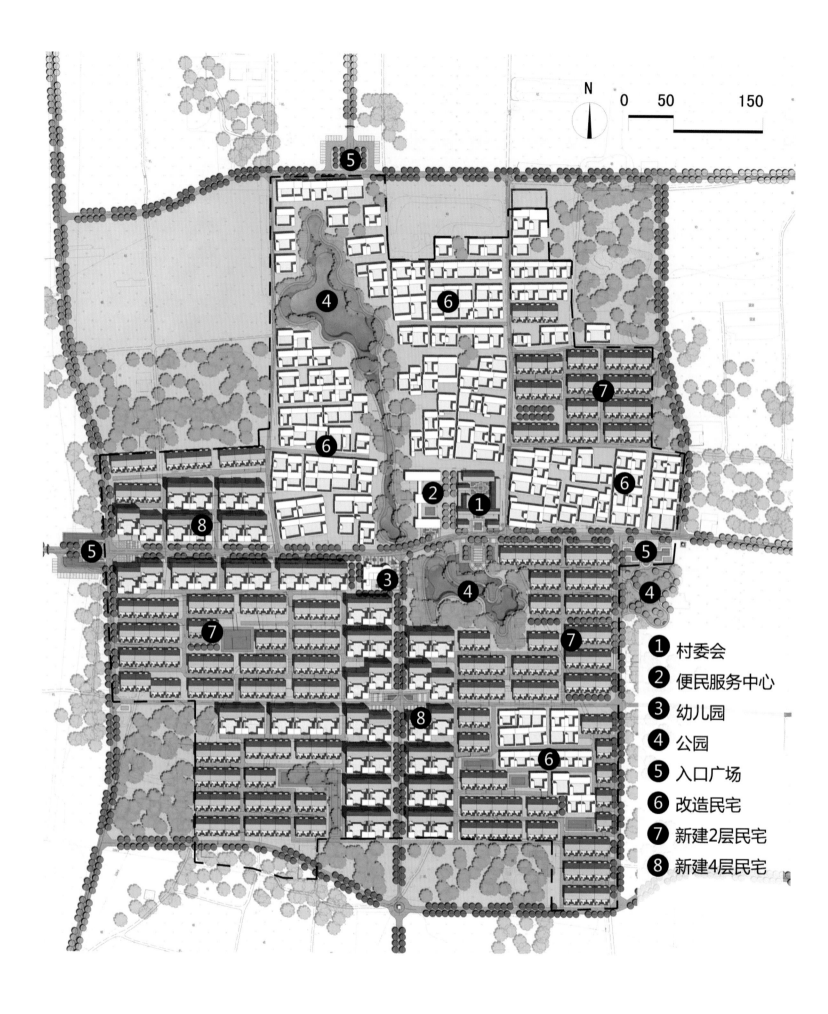

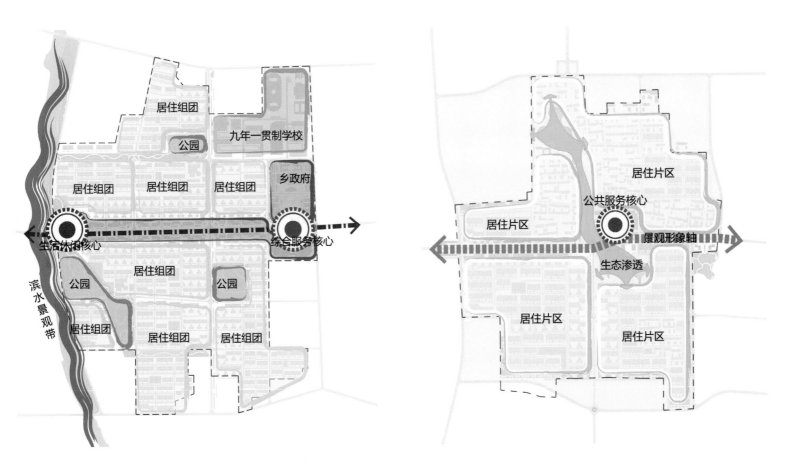

THE PLANNING AND DESIGN OF THE NEW YUHONG INDUSTRIAL TOWN.

于洪产业新城

Location

辽宁省沈阳市
Shenyang City, Liaoning Province

Area

约 0.71 平方千米
About 0.71 square kilometers

Time

2013 年至今
Since 2013

"文化社区"的概念

以简欧风格建筑作为文化切入点，在造型和组团空间组织上形成强有力的文化标签，使住区形成高度的可识别性。

住区内部采用西方庭院的造景手法营造绿化景观，设置多处文化广场和人文主题雕塑，使建筑和景观浑然一体，具有高度的文化统一性。

在住区中心设置图书室，为住户提供阅读和文化交流的场所，努力打造"文化社区"，实现全民阅读，点亮中国梦。

"借远景、框中景、纳近景"的策略

规划借蒲河远景，框轴线中景，纳花园近景，充分考虑人的感受，让建筑与环境共生。简约、古典的建筑呈现出静谧高雅的情调。

"CULTURAL COMMUNITY" CONCEPT

This is to form a strong cultural identity in styling and spatial arrangement, by using simple European style architecture as a cultural inspiration, so as to ensure a high degree of residential identity.

In the community, the style of western garden will be adopted to foster green landscapes, and set up multiple cultural plazas and cultural theme sculptures to guarantee the harmony of architecture and landscape, achieving a high degree of cultural unity.

A library will be located in the community center, a place for reading and cultural exchange, which is a means to create a "cultural community", getting people engaged in universal reading and lighting up the Chinese dream.

"BORROWING THE VISTA, FRAMING THE MEDIUM SCENE, ACCESSING THE CLOSE-RANGE SCENE" STRATEGY

The planning is to make good use of distant scenery of Puhe River, frame the medium scene — the axis of the downtown, and render the close-by garden accessible to the residents of the community. This design aims to create a symbiotic relationship between architecture and nature, so that the residents will have a sense of oneness with their surroundings. This creates a quiet and elegant atmosphere, simple in style and classic in architecture characteristics.

THE PLANNING AND DESIGN OF THE NEW TOWN IN NORTHERN GU'AN COUNTY

固安县北部新城规划设计

Location
河北省廊坊市
Langfang City, Hebei Province

Area
0.61 平方千米
0.61 square kilometers

Time
2014 年至今
Since 2014

固安县隶属河北省廊坊市，地处华北平原北部，北京、天津、保定三市中心区域。伴随着社会和经济的发展，固安的居住环境也期待着进一步的提升。

打造庄重、现代的特色新城

相应地提出五大策略：

策略一：合宜的邻里架构。

策略二：完善的配套设施。

策略三：多元的公共空间。

策略四：古典的空间轴线。

策略五：庄重的建筑风貌。

Gu'an County is under the jurisdiction of Langfang City, Hebei Province, located in the north of North China Plain and the center of the three cities, Beijing, Tianjin, and Baoding. With the development of society and economy, the living environment of Gu'an is also expected for further promotion.

TO CREATE A FEATURED NEW TOWN WITH SOLEMN AND MODERN CHARACTERISTICS

Accordingly, five strategies are proposed:

Strategy 1: a proper neighborhood structure.

Strategy 2: perfect supporting facilities.

Strategy 3: multiple public spaces.

Strategy 4: a classical spatial axis.

Strategy 5: a solemn architectural style.

THE PLANNING AND DESIGN OF THE NEW HUAXIA TOWN IN YONGQING COUNTY

永清县华夏新城规划设计

Location	Area	Time
河北省廊坊市 Langfang City, Hebei Province	约 18 平方千米 About 18 square kilometers	2013 年 2013

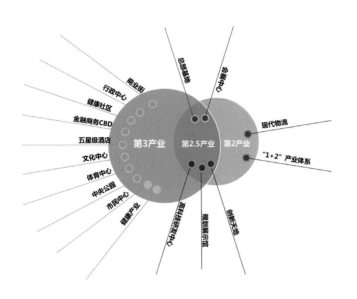

如今，环京地区正在寻找各自的突破方向，廊坊市永清县也面临着机遇与挑战。

产业定位：为了绿色健康的生活

华夏新城的设计定位于打造全球影视高新技术实践区，规划概念如下：

绿动基础——以水道为骨，让绿色游走。

健康产业——平台支撑，全域健康。

新鲜生活——森鲜供应，亲水乐活。

双核驱动，绿色脉搏

项目格局可以确定为四大定位：绿色生态之城、美好活力之城、开放共享之城、智慧创新之城。

基于综合条件，提出如下方案：

方案一：双核引领，板块共生，蓝网绿脉，多元水岸。

方案二：绿色纽带，城脉（水系）贯通，网状渗透，生态营城，双核引领，均衡发展。

方案三：生态楔入，绿色引领，双心联动，轴带发展。

Today, the Greater Beijing Region is looking for the direction of the regional breakthrough, and Yongqing County,Langfang City faces both opportunities and challenges.

ORIENTATION OF THE INDUSTRY: FOR GREEN AND HEALTHY LIFE

The design of the new Huaxia Town is orientated as the high-tech practice area of global film and television, and has the following concepts:

Foundation of "moving green": using the waterway as the bone to let the green meander along.

Health industry: supported by the platform in achieving universal health.

"Fresh" life: with the supply of "the fresh", to have a happy life on the waterfront.

DUAL-CORE DRIVE, PULSATION OF THE GREEN

The pattern of the project can be identified as having four major orientations: the city of green and ecology,the city of beauty and vitality, the city of open and sharing, the city of wisdom and innovation.

Based on the overall conditions, three schemes are put forward:

Scheme 1: the leading role of dual-core drive, symbiosis between different blocks, "blue" network and "green" pulse, and multiple waterfronts .

Scheme 2: the green linkage, city pulse (water system) connection, network penetration, eco city, the leading role of dual-core drive, and balanced development.

Scheme 3: ecological interaction, the leading role of the green, dual-core integration, and developing in the pattern of axis and belt.

方案一
双核引领，板块共生，蓝网绿脉，多元水岸

方案二
绿色纽带，城脉（水系）贯通，网状渗透，生态营城，双核引领，均衡发展

方案三
生态楔入，绿色引领，双心联动，轴带发展

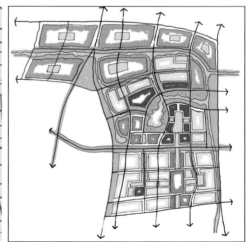

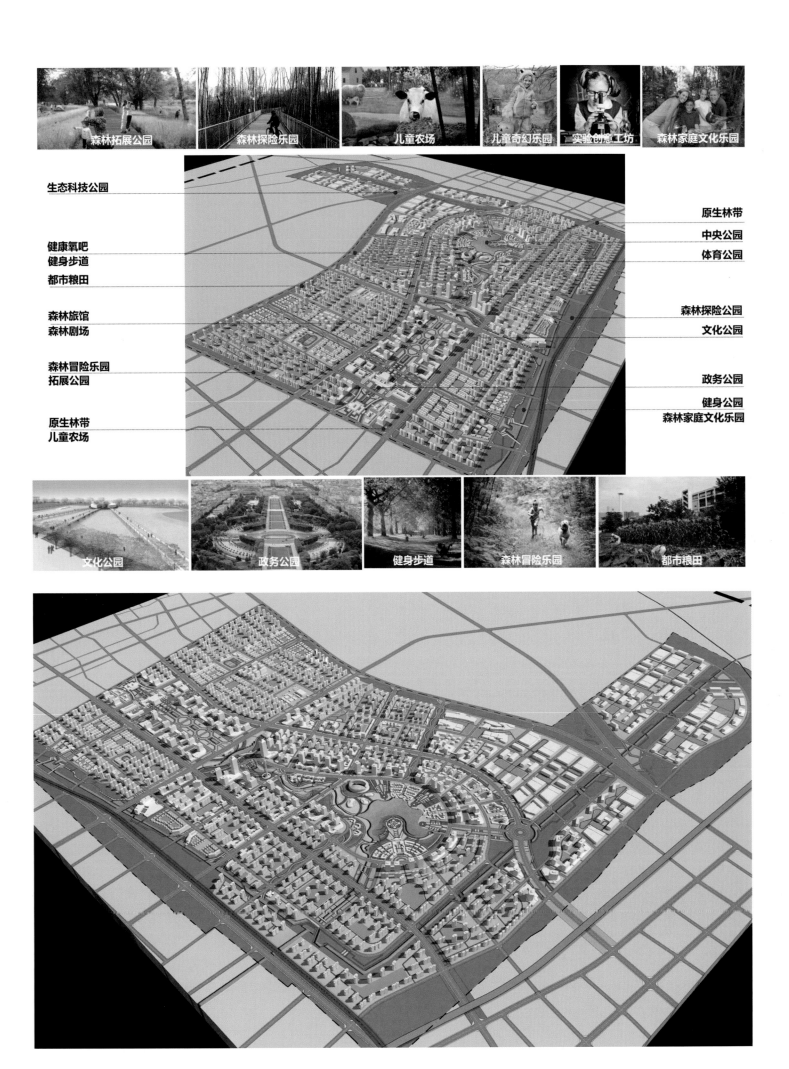

THE PLANNING AND DESIGN OF THE NEW TOWN IN GU'AN COUNTY

固安县新城规划设计

Location

河北省廊坊市
Langfang City, Hebei Province

Area

5.59 平方千米
5.59 square kilometers

Time

2013 年
2013

固安新兴产业示范园区地处北京中轴线南端，距离天安门广场50千米，与北京隔永定河相望，处于大北京经济圈的前沿部位。

园区距离北京大兴国际机场仅10千米，距北京南三环39千米，距天津110千米。驱车1小时可以到达北京首都国际机场，1.5小时可以到达天津港。

创建复合型产业新园

核心区将被打造成一个由商务、科研、商业、文化、休闲、居住等区域组成的国际化、复合型的产业园核心区，树立产业园的新形象，提升产业园的知名度。示范区将立足固安，连接京津，面向全国，打造国际化的工作、生活、休闲聚集地。

融汇商务、产业与文化生活

规划提出这样的愿景：

最具地标的商务科研集聚地：为固安新兴产业示范园区打造一个独特的现代商务科研形象。

最具智慧的产业孵化基地：提供智能化、低密度、生态型的总部楼群。

最具活力的商业文娱综合体：缔造24小时多元化工作和生活方式融合的典范。

最具品质的生态宜居目的地：提供适宜居住的环境，引入高端的物业，创造精致的生活品质。

最具智能的城市可运营系统：覆盖全区的无线网络平台，城市居民可随时随地便捷地连接网络。智能的交通设施，发达的物联网生活平台，以及围绕智能城市可运营的系统。

Gu'an Emerging Industry Demonstration Park is located at the southern end of the Central Axis of Beijing, and 50 kilometers away from Tian'anmen Square. It is opposite to Beijing across the Yongding River and situated in the front part of the Big Beijing Economic Circle.

The park is only 10 kilometers away from Beijing Daxing International Airport, 39 kilometers away from Beijing South Third Ring Road, and 110 kilometers away from Tianjin. It takes an hour to drive to Beijing Capital International Airport and 1.5 hours to drive to Tianjin Harbor.

BUILDING A NEW INTEGRATED INDUSTRIAL PARK

The Core Park will be the core of an integrated international industrial park, consisting of business, scientific research, commerce, culture, leisure and residence area, to establish a new image of the industrial park and increase its reputation. The Demonstration Park will consolidate its position in Gu'an to connect Beijing and Tianjin, and be built into an international congregation for work, living and leisure for the whole nation.

INTEGRATING BUSINESS, INDUSTRY AND CULTURAL LIFE

The prospects of the project are:

A landmark for business and scientific research congregation: to build a unique image of modern business and scientific research for the Emerging Industry Demonstration Park of Gu'an.

The most smart industry incubation base: to provide intelligent, low-density and ecological headquarters buildings.

The most vigorous commerce, culture and entertainment complex: to establish integration model of 24-hour diversified work and lifestyles.

The ecological destination with the highest quality of living: to provide livable environment, high-level property management and exquisite life quality.

The most intelligent city operational system: available Wi-Fi for urban residents at anytime and anywhere, smart transportation facilities, advanced life necessities platform for the Internet of Things, and an operational system for intelligent city.

THE PLANNING AND DESIGN OF THE NEW AIRPORT ELITE CITY

新空港精英城市规划设计

Location

河北省廊坊市
Langfang City, Hebei Province

Area

6.31 平方千米
6.31 square kilometers

Time

2012 年至今
Since 2012

新空港精英城市是固安产业新城的重要一极。它不可脱离城市化进程而独立存在，必须充分整合资源，与产业、城市共同发力，才能实现开发价值的最大化。

缔造强大的"中心"

规划将致力于为小镇打造四个中心：

精神地标——小镇中心首先是整个社区的精神中心。

商业中心——小镇中心其次是整个社区的商业配套集中地。

形象标签——小镇中心还是整个社区的形象展示窗口。

生活舞台——小镇中心最后是整个社区的人群与活动聚集地。

The New Airport Elite City is an important part of Gu'an New Industrial Town, which can not exist independently from the urbanization process, and must fully integrate resources and work together with the industry and city, so as to maximize the value of development.

BUILDING STRONG "CENTERS"

The plan will be dedicated to creating four centers for the small town:

Spiritual landmark – the town center is first and foremost the spiritual center of the community.

Business center – the town center secondly congregates the commercial facilities of the whole community.

Identity symbol – the town center is also a window for displaying the image of the whole community.

Stage of life – last but not least, the town center is a place where people of the whole community gather and have social activities.

城市级配套项目
1. 体育公园
2. 儿童组合运动场、儿童酷骑自行车运动场
3. 大操场
4. 森林公园
5. 家庭休闲公园
6. 都市沙滩、儿童高尔夫
7. 矮马公园、撒野公园、烧烤草坪
8. 农耕公园
9. 动植物公园
10. 自然公园
11. 儿童交通规则公园

1. 国际小学
2. 国际中学
■ 大超市

社区级配套项目
1. 农耕公园
2. 动植物公园
3. 自然公园
4. 儿童交通规则公园
5. 垂钓园、草地保龄球场
6. 健身运动公园
7. 阳光玻璃花房、艺术廊道、风雨连廊、家庭文化乐园

8. 儿童组合运动场、儿童酷骑自行车运动场
9. 都市沙滩、儿童高尔夫
10. 矮马公园、撒野公园、烧烤草坪
11. 药膳公园
12. 慢行系统

1. 培训机构、儿童剧场、DIY创意工坊、邻里之家、画廊旅馆
2. 幸福图书馆、书店、儿童书店
❶ 老年大学、老年医院
❷ 老年俱乐部、老年健康护理中心
❸ 老年餐厅、老年护理培训中心
❹ 社区医疗中心

5. 儿童护理中心、酷骑运动馆
6. SPA养生馆、运动场馆、健身养生会所
7. 药膳餐厅
1. 商务会所、社区风情商业街
2. 精英俱乐部、啤酒吧、咖啡吧

组团级配套项目
▲ 幸福会所
△ 组团绿地

THE PLANNING AND DESIGN OF THE NEW SOUTHERN TOWN IN GU'AN COUNTY

固安县南部新城规划设计

Location	Area	Time
河北省廊坊市 Langfang City, Hebei Province	7.39 平方千米 7.39 square kilometers	2012 年至今 Since 2012

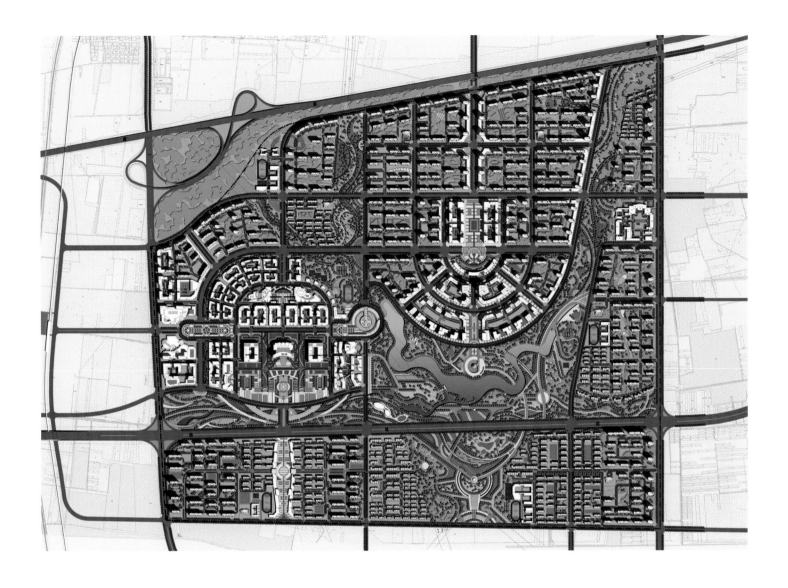

所谓幸福,就是在较长时间内对生活感到满足,认为生活中有巨大的乐趣并自然而然地、持续地拥有愉快的心情。这一规划的目标就是打造一座"幸福新城",一座和谐、健康、快乐的低碳新城。

"一环两轴,一核三心"

生态绿环——依托现有绿化格局形成生态绿网,将组团节点与绿网结合,点轴辉映形成区域功能框架。

新区发展核——规划依托区域规划及现状条件,形成区域发展核心,涵盖行政办公、商务会展、文化娱乐等多种综合性功能。

The so-called happiness means people's continuous satisfaction with life over a comparatively long period. They think they can get tremendous delight from life, and naturally, maintain a good mood for long. The mission of this plan is to build a "New Happy Town" where there are harmony, health, happiness and low carbon emission.

"ONE RING AND TWO AXES, ONE CORE AND THREE CENTERS"

Eco-green ring— it depends on the existing green pattern to develop an eco-green network, which will combine group nodes with the green network, with points and axes coordinating into a regional functional framework.

The new town development core — it relies on the regional planning and current conditions to form a regional development core with multiple comprehensive functions including administrative offices, business exhibitions and cultural entertainments.

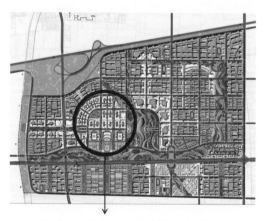

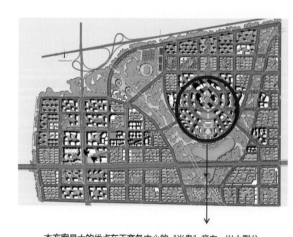

本方案最大的优点在于行政中心充分考虑中国传统文化和空间特色，同时借鉴了西方立法、行政和司法三权分立的思想，打造一个独具地域特色的行政中心空间形象。

本方案最大的优点在于商务中心的"半岛"意向，以大型公园环绕现代化商务中心，展示出了恢弘大气的开放空间。

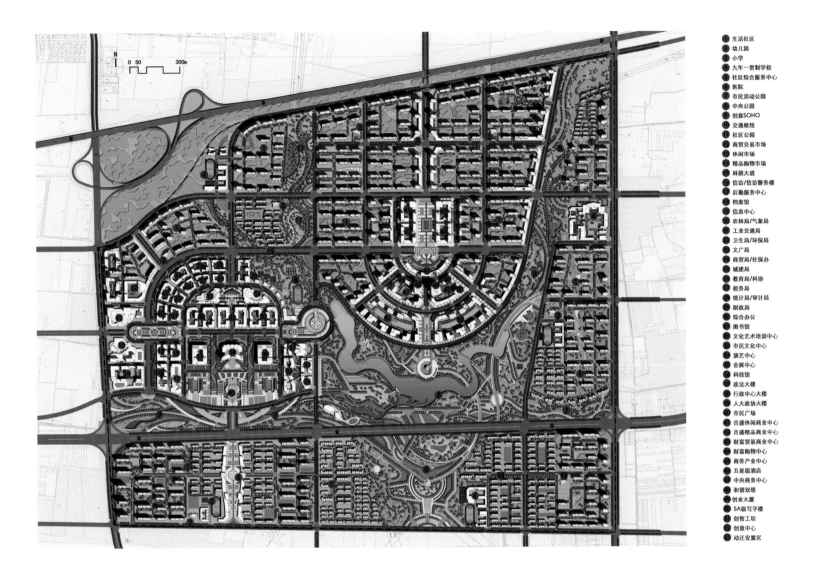

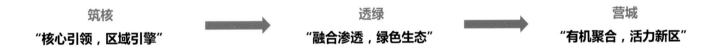

筑核	透绿	营城
"核心引领，区域引擎"	"融合渗透，绿色生态"	"有机聚合，活力新区"

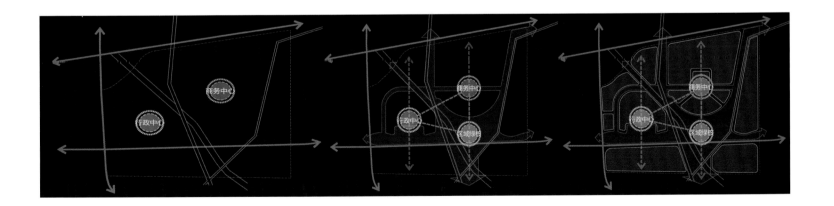

THE PLANNING AND DESIGN OF THE NEW TOURIST TOWN OF GOLDEN COAST

金渤海岸旅游新城规划设计

Location
辽宁省大连市
Dalian City, Liaoning Province

Area
约 24.05 平方千米
About 24.05 square kilometers

Time
2013 年至今
Since 2013

规划的目标是依托海洋资源，打造一个集总部经济、滨海时尚休闲、海景度假养生、湿地公园、生态居住为一体的东北亚国际旅游度假区。依托机场经济，打造一个兼具机场物流、综合商贸和总部经济功能的东北亚空港经济区。

金州秘境，钟鸣四海

"金州"指金州湾，"秘境"指金州湾多近海的岛屿，堪称离岸秘境，岛屿形态万千，形成金州湾的主体资源。"钟鸣"源于"钟鸣鼎食"，"四海"则源于"富有四海"，寓意金州湾通过自己不断的发展，吸引有识之士，实现文化、科技与财富的积累。

和谐共生

规划采用有机聚合的布局模式，以水为脉构建水脉交融的生态网络，各片区向心集聚发展，并通过打造城市主干道路，强化各片区的联系。

规划布局为"一湾、二核、三中心"。

金螺湾以海螺形环抱式空间寓意拥抱未来，高低错落的中低密度商贸建筑群将成为区域发展的智慧引擎。

"二核、三中心"以海岸为主要线性发展空间，串联度假、产业等两大核心，辅以三大次级中心，由点及面全面触动区域活力。

打造"盈彩水岸连五洲"的精彩布局。

金螺岛（会务岛）是集高端商贸、花园总部、会议论坛、创意中心为一体的区域活力核心，打造滨海财富金岸。红礁岛（度假岛）以滨海休闲、养生度假、运动疗养为主题，依托生态特色，打造隐于滨海地区的度假养生地带。星鲨岛（文娱岛）以国际文化娱乐引领城市生活，打造具有国际影响力的浪漫文化之岛。蝶鱼岛（起航岛）以物流产业为核心动力，打造集研发、办公、休闲为一体的综合生态产业基地。雀鲷洲（宜居洲）是以水岸生活为主题的养生高尚社区。

The plan, taking advantages of the marine resources, aims to build an international tourist resort in Northeast Asia, integrating headquarters economy, coastal fashion and leisure, coastal landscape resort and regimen, wetland park as well as ecological living. Making use of the airport economy, there will be an Airport Economic Zone in Northeast Asia with the functions of airport logistics, comprehensive commerce and headquarters economy.

SECRET PLACE IN JINZHOU, BELL RINGING OVER THE FOUR SEAS

"Jinzhou" refers to Jinzhou Gulf; "secret place" means various coastal islands and islets which are quite cut off but close to the mainland, featuring different shapes, forming the main resources. "Bell ringing" means extravagance of life. "The four seas" mean great prosperity, symbolizing the realization of the sustainable accumulation of culture, science and technology, and wealth through continual development and attraction to intelligent and wise people.

HARMONIOUS SYMBIOSIS

The plan adopts the layout mode of organic convergence, where waterways are used as the vein to create an eco network, and each area gathers and develops concentrically to the core, meanwhile enhances the connections of all the areas by building the main trunk roads of the city.

The layout of the plan is "one gulf, two cores and three centers".

The space of Golden Conch Gulf that features the shape of conch rings symbolizes embracing the future. The high-and-low zigzag commercial buildings with medium-and-low density will become the wisdom engine in regional development.

"Two cores and three centers" mean that the coast will be the principal linear development space, connecting the two major cores of resort and industry, and is to be supported by three sub-centers, fully triggering the regional vitality from a point to a big area.

The plan creates a wonderful layout of "colorful coast connecting five continents".

Golden Conch Island (for conference) is the regional vitality center, to build a golden coast of wealth, including high-level commerce, garden headquarters, conference forum and innovation center. Hongjiao Island (for resort) is themed by coastal leisure, regimen

holiday and sports rehabilitation, aiming to build a hermit like holiday resort based on the ecological feature. Xingsha Island (for entertainment) will be constructed into an internationally influential romantic cultural island, leading the urban life by international cultural entertainment. Dieyu Island (for transportation), taking the logistics industry as the core dynamic, will be built into a comprehensive ecological industrial base integrating research and development, office work and leisure. Quediao Island (for livability) is a high-level regimen community themed by the coastal life.

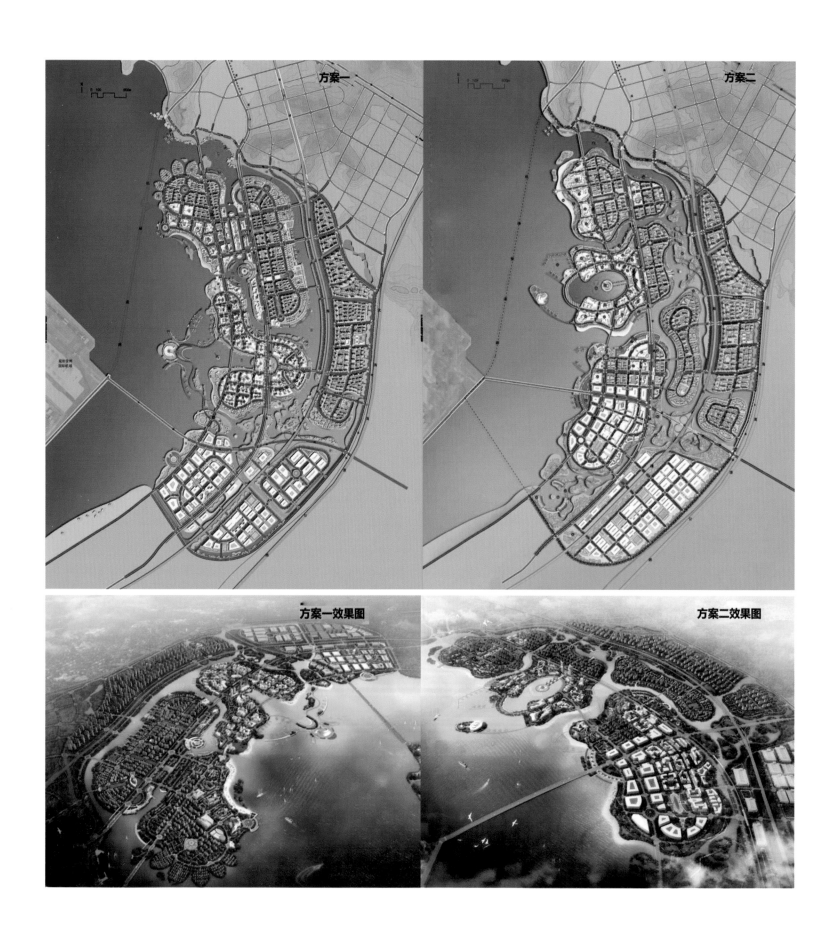

方案一

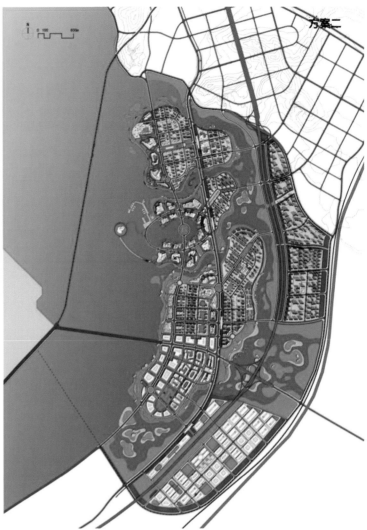

方案二

方案一三维图

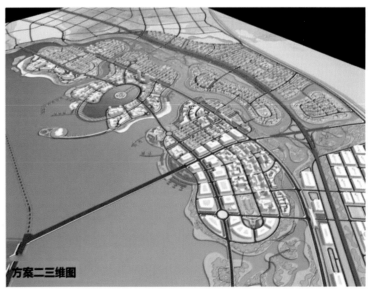

方案二三维图

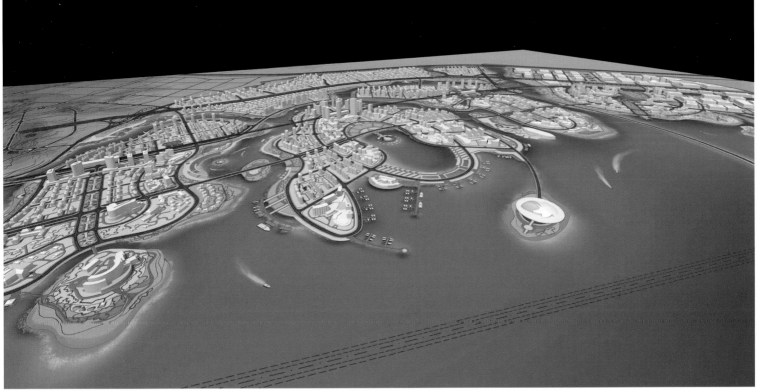

THE PLANNING AND DESIGN OF THE TOURIST TOWN OF JINYUE GULF IN LINGSHUI

陵水金月湾旅游小镇规划设计

Location

海南省东方市
Dongfang City, Hainan Province

Area

约 13.95 平方千米
About 13.95 square kilometers

Time

2013 年至今
Since 2013

打造高端康养小镇

海南旅游业的发达，给予项目优越的开发背景。

规划主张生态优先，功能支撑，立足海南旅游圈，依托南海、尖峰岭森林公园的生态自然资源，秉承黎族、苗族等特色人文资源，打造一个拥有优质沙滩、提供国际化游乐体验的高端康养圣地，打造综合生态休闲、文化体验、康体养生和论坛度假等多元特色，国内领先、全球知名的国际高端综合休闲度假区。

CREATING A HIGH-END HEALTH RESORT

Hainan's tourism industry is well-developed, providing a surperior development background for this project.

The planning, taking ecology as a priority and function as support, will make good use of the Hainan tourism circle to build a high-end health and regimen resort with high quality beaches and international entertainment, on the basis of the ecological and natural resources of the Nanhai Sea and Jianfengling Forest Park, as well as special human resources of Li and Miao ethnic groups. A high-end international resort for leisure and holiday will be built to be the leader within the country, renown all over the world through the integration of multiple functions of ecological leisure, culture experience, health and regimen, forum and vocation, etc.

THE PLANNING AND DESIGN OF THE WEST EUROPEAN CULTURAL TOWN IN NINE-LOONG CANYON, HENGDIAN TOWN

横店镇九龙大峡谷西欧风情小镇规划设计

Location	Area	Time
浙江省东阳市 Dongyang City, Zhejiang Province	约 0.27 平方千米 About 0.27 square kilometers	2013 年至今 Since 2013

九龙水多、坡多，山水相映，动静相宜，刚柔相济，山水一色。这片柔媚的山间水境，是休闲疗养的好去处。九龙正向世人敞开她热情的胸怀，走进九龙，就是投入了自然的怀抱。

西欧风情，生态度假

规划汇集欧洲经典建筑风情，以休闲疗养度假为特色，集高端养老、康复疗养、生态度假、运动休闲、文化艺术、SPA养生于一体，兼具婚庆服务与影视拍摄场地，打造山水秘境、异域浪漫的九龙大峡谷度假酒店群落。

The Nine-Loong Canyon has plenty of rivers and slopes, and the water system and mountains correspond to each other in motion and tranquility, in strength and gentleness, forming a harmonious whole. This piece of beautiful place with water system and mountains constitutes a perfect resort for relaxation and rehabilitation. Nine-Loong Canyon extends her warm arms to welcome all the people. Coming to Nine-Loong is coming to the embrace of Mother Nature.

WEST EUROPEAN STYLE, ECOLOGICAL VACATION

The program is to have a collection of the styles of European classic architectures, and features an integration of high-end service for the elderly, rehabilitation healthcare, ecological vacation, sports and leisure, culture and arts, and SPA service, meanwhile provides a place for wedding ceremonies and TV screenings, creating a cluster of romantic Nine-Loong Canyon holiday hotels for an escape to watery mount with exotic taste.

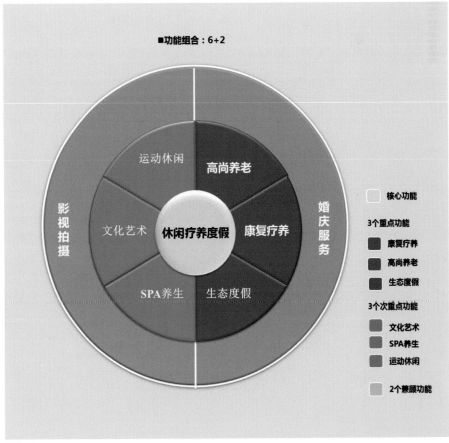

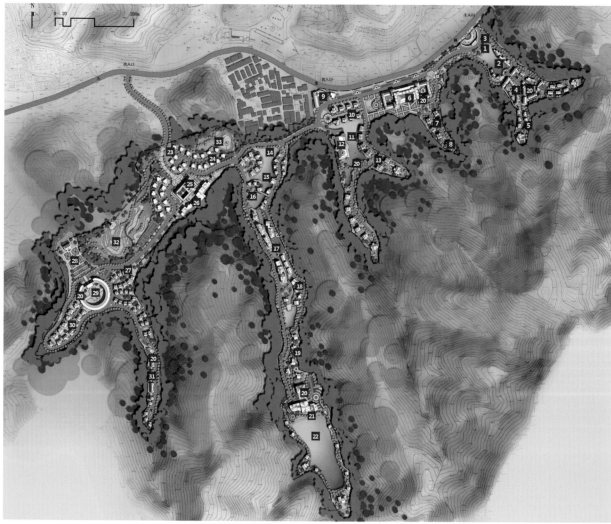

163

THE PLANNING AND DESIGN OF THE INTERNATIONAL JEWELRY CULTURE INDUSTRIAL TOWN

国际珠宝文化产业小镇规划设计

Location

云南省瑞丽市
Ruili City, Yunnan Province

Area
0.16 平方千米
0.16 square kilometers

Time
2013 年至今
Since 2013

依托原料市场和终端市场，打造珠宝翡翠的物流、贸易平台。以物流仓储、商务贸易为重点产业，以旅游服务、文化展示为辅助产业。主体项目包括珠宝翡翠保税仓库、公盘交易标场、国际珠宝城（终端零售）。

产业推动：珠宝翡翠加工

弥补产业薄弱环节，打造珠宝翡翠的加工聚集地。以加工为重点产业，兼顾旅游服务、文化展示和贸易。主体项目包括翡翠加工展示中心、翡翠加工综合街区、翡翠加工LOFT、翡翠加工创业孵化园。辅助项目包括海关办公中心、人才培训中心和酒店式公寓。

创意推动：珠宝设计高地

发展大师级艺术家群落，打造珠宝翡翠的创意设计高地。以珠宝设计为重点产业，兼顾旅游服务、文化展示和贸易。主体项目为创意中心（臻品会所）。

旅游推动：兼顾文化展示与珠宝贸易

传播珠宝翡翠文化，打造面向东南亚的形象展示窗口。以文化展示为重点产业，兼顾旅游服务和贸易。主体项目为国际珠宝会展中心（包括展示区、博物馆、拍卖中心等）。

加强旅游服务配套，立足瑞丽珠宝翡翠文化产业，打造具有当地特色风情的玉石加工产业型旅游目的地。以旅游服务为重点产业，兼顾文化展示和贸易。主体项目为配套娱乐、餐饮新天地。

Relying on raw materials market and the terminal market, it is aimed to create a logistics and trade platform for jewelry and jade. Giving priority to logistics warehousing and business trade, tourism services and cultural display are undertaken as auxiliary industries. The main projects include jewelry and emerald bonded warehouses, the public trading bidding site and the international jewelry city (terminal retail).

INDUSTRY PROMOTION: JEWELRY AND EMERALD PROCESSING

It will make up for the weak sectors of the industry, and create a processing gathering place on jewelry and jade. Taking processing as a key industry, the related industries are taken into account, such as tourism services, cultural display and trade. The main projects include emerald processing exhibition center, emerald processing integrated block, emerald processing loft, Incubation Park of emerald processing entrepreneurship. Auxiliary projects include customs office, talent training center and hotel-style apartments.

CREATIVE PROMOTION: JEWELRY DESIGN HEIGHT

The project will develop the master artist community and create a creative design height of jewelry and jade. Counting design as a key industry, others such as tourism services, cultural display and trade are to be developed as well. The main project is the creative center (Masterpiece Club).

TOURISM PROMOTION: COVERING BOTH CULTURAL DISPLAY AND JEWELRY TRADE

The design aims at spreading jewelry and jade culture and creating an image display window to Southeast Asia. The focus is on cultural industries, but tourism services and trade are not to be neglected. The main project is the international jewelry convention and exhibition center (including an exhibition area, a museum, an auction center, etc.).

By strengthening the supporting facilities of tourism service, Ruili City will take a foothold on jewelry and jade cultural industry, to create a jade processing industry-oriented tourism destination with local characteristics. Apart from tourism services as the key, the cultural display and trade are to be cultivated, too. The main projects here are the supporting entertainment facilities and new dining ground.

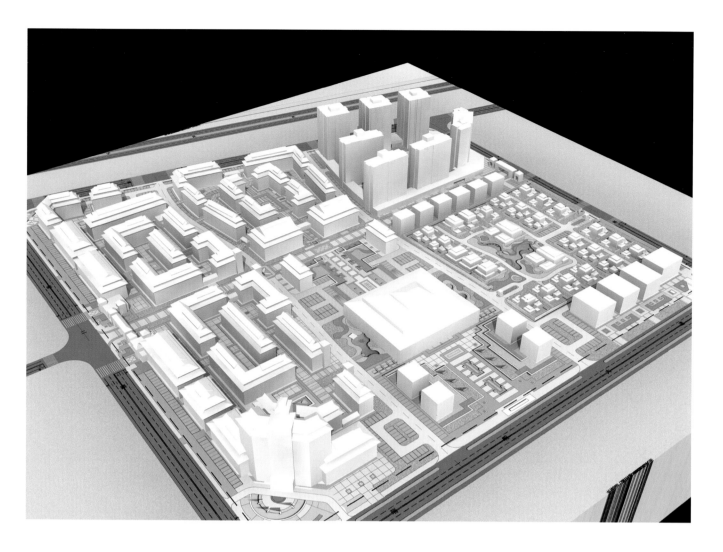

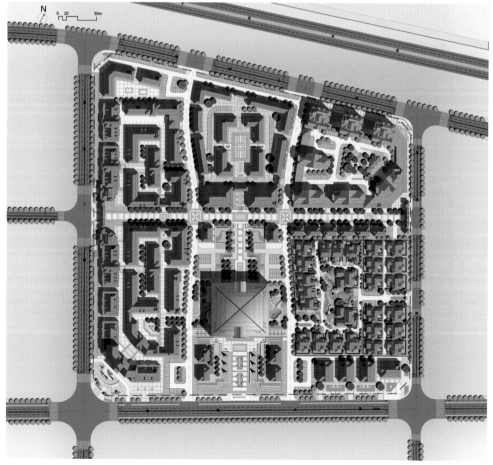

1. 企业孵化器
2. 珠宝职业培训中心
3. 珠宝创意工坊
4. 珠宝贸易中心
5. 海关办公
6. 珠宝加工中心
7. 保税仓库及公盘交易标场
8. 免税中心
9. 品牌珠宝专卖店
10. 翡翠时尚广场
11. 翡翠文化广场
12. 国际珠宝会展中心
13. 珠宝城
14. 企业总部基地
15. 加工SOHO
16. 加工LOFT
17. 酒店

THE PLANNING AND DESIGN OF THE NEW SOUTHEAST ASIAN AND SOUTHERN ASIAN CULTURE AND COMMERCIAL TOWN IN DONGBA

北京市东坝东南亚、南亚文化商贸新城规划设计

Location

北京东坝国际商贸中心区
International Trade Center District, Dongba, Beijing

Area
0.13 平方千米
0.13 square kilometers

Time
2013 年至今
Since 2013

汇聚异域风情文化与特色商品

东坝国际商贸中心区，是一片充满机会的热土。

京城异域风情汇集之地——汇集东南亚、南亚各国风情建筑，极具识别性的标志建筑之地。

国际特色商品交易之所——汇集东南亚、南亚各地特色商品，千姿百态、色彩斑斓的交易之所。

特色民族文化汇聚之地——东南亚、南亚各民族文化提萃，多元文化交融的活力展示之窗。

打造充满魅力的文化商贸综合体

项目提出六大开发策略，大力打造集文化演艺、餐饮娱乐、批发零售、商业购物、精品酒店、商务办公与配套公寓于一体的文化商贸综合体。

实现功能集聚，形成产业高地。通过延展、提升基地内部功能的基本门类，展现其商业、商务、贸易、办公、餐饮、娱乐的独特面，形成复合型贸易集聚区。

创造多元体验，营造特色街区。汇聚东南亚、南亚多种建筑风格与空间特色，打造极具地域标志性的城市空间。

聚焦文化提升，彰显地域风貌。通过多种地域文化、休闲文化等多元文化注入，实现项目多元提升。

BRINGING TOGETHER EXOTIC CULTURES AND SPECIALTY GOODS

Dongba International Trade Center District is a place full of chances.

Land of exotic cultures in Beijing: a collection of amorous buildings from the Southeast Asian and Southern Asian countries, a place with outstanding identity of landmark architectures.

A place of international commodity trade: a collection of characteristic products from various places of the Southeast Asian and South Asian countries, a trading place of wonderful expressions of colorful goods.

A place of characteristic ethnic cultures: a collection of various ethnic cultures of the Southeast Asian and South Asian countries, an exhibition of the vigor of multi-cultural blending.

TO CREATE A CHARMING CULTURAL AND COMMERCIAL COMPLEX

The project puts forward six development strategies, to create a cultural and commercial complex featuring cultural performance art, restaurants and entertainment, wholesale and retail, commercial shopping, the Inn Boutique, business office and supporting apartment facilities.

This is to achieve functional agglomeration, and to create industrial heights. It is to create a complex trading gathering area with special features in business, business service, trade, office, catering, and entertainment, through extension and upgrading the basic categories of the internal functions of the base.

It will create diverse experiences and blocks with special features. Converging various architecture styles and space features from the Southeast Asia and Southern Asia, it will create an urban space with impressive regional landmark.

It is to focus on promoting culture and highlighting diverse geographical features. By integrating the multiple regional cultures, as well as the leisure culture, it will realize the objective of upgrading multiple functions of the project.

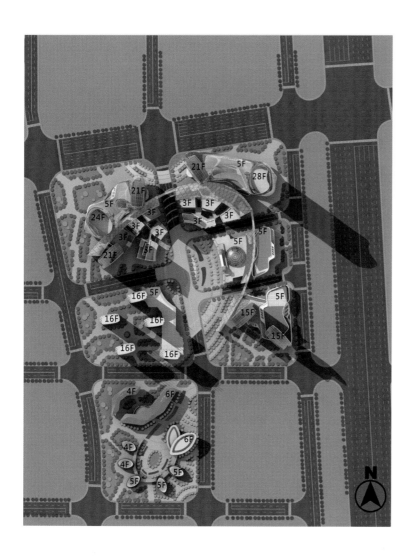

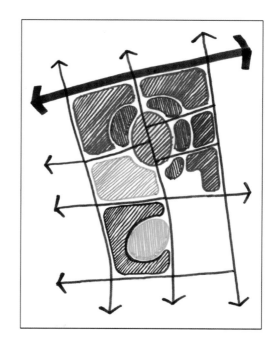

总体技术经济指标		
用地面积		128200m²
总地上建筑面积（计容）		401230m²
其中	商业面积	119830m²
	酒店面积	19100m²
	办公面积	203600m²
	公寓面积	51000m²
	剧场面积	7700m²
建筑密度		42.7%
容积率		3.13
绿地率		30.0%

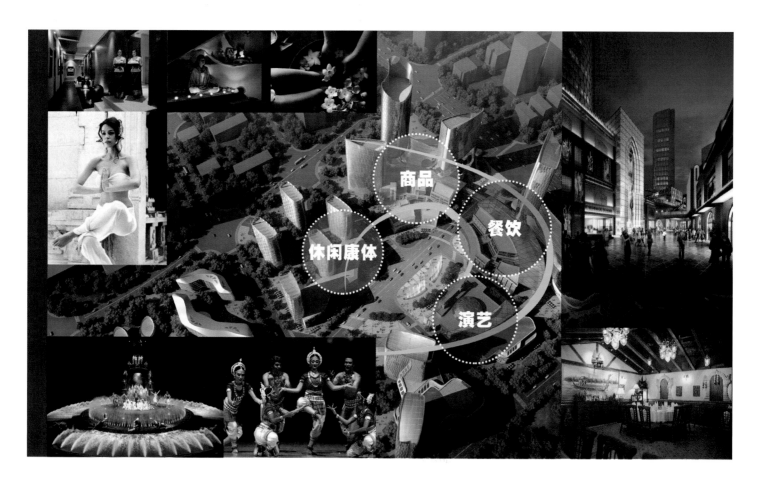

THE PLANNING AND DESIGN OF THE WATERY TOWN OF SOUTHERN CHINA, HENGDIAN TOWN

横店镇江南水乡风情小镇规划设计

Location

浙江省东阳市
Dongyang City, Zhejiang Province

Area

约 0.12 平方千米
About 0.12 square kilometers

Time

2013 年至今
Since 2013

以电影情节体验为主题

项目有着明确的品牌定义,那就是"以城市和文化的名义",致力于打造中国标杆性的、以电影产业为背景、具有特色文化元素、以电影情节体验为主题的城市中心体验式商业街区。

项目的市场定位在于以餐饮、休闲、旅游、娱乐为着力点,形成富有特色及吸引力的综合旅游、饮食、娱乐、休闲文化于一体的商业文化街区。

水乡里的"中国好莱坞"

规划将为横店影视城打造一个"中国好莱坞"的城市名片,以江南水乡风情与横店古建筑、"中国好莱坞"文化事件为主题吸引国内外游客。

项目确立了自身的功能定位,即以周边特色景区为载体的休闲区、以旅游休闲娱乐为主导的商业区、以影视文化创意为特色的体验区。

横店江南水乡商业街将以文化产业为核心,发展以文化为中心进行延展的创意产业。同时,充分考虑资源禀赋、区位优势、产业基础和区域分工协作等因素,以旅游服务产业为辅助产业,基于核心产业,确定横店江南水乡商业街的产业组合。

THEMED ON EXPERIENCING MOVIE SCENARIOS

The project is featured with a clear brand definition, which means that it is "in the name of city and culture" to be committed to creating a Chinese bench-marking central city experience business district film industry as the background, culture as its characteristics, and movie scenarios experiencing as the theme.

The market orientation of the project is to start with catering, leisure, tourism, and entertainment, to create a unique and attractive culture street, integrating touring, restaurants, entertainment, and leisure culture.

"CHINESE HOLLYWOOD" IN A WATERY TOWN

The plan is to build a city card of "Chinese Hollywood" for Hengdian World Studio to attract tourists home and abroad with the combination of Southern China watery town style and Hengdian ancient buildings, and with the theme of "China Hollywood" cultural events.

The project establishes its functional orientation as to create a district of experience with a leisure area by making use of the surrounding scenic spots as the carrier, a business district focusing on tourism and entertainment, and an experiencing area featuring film and television cultural creations.

Hengdian Southern China watery town business district is to focus on cultural industry, to develop extended creative industry centering on culture. Furthermore, by taking full account of the factors like resource advantages, location advantages, industrial base and regional division of labors, and based on the core industries, with the support of tourism and service industries, the project is to establish the industrial agglomeration of Hengdian Southern China watery town business district.

THE PLANNING AND DESIGN OF THE NEW TOURIST TOWN WITH THE HISTORY AND CULTURE OF THE THREE KINGDOMS

三国历史文化旅游新城规划设计

Location	Area
云南省曲靖市 Qujing City, Yunnan Province	约 2.15 平方千米 About 2.15 square kilometers

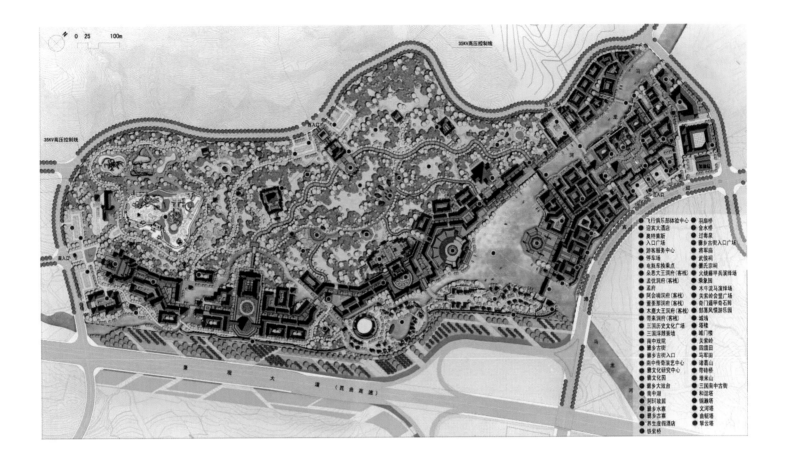

规划以"风中云南,融和曲靖"为主题,以三国时期故事、爨文化为主,兼容地方文化要素,形成可供游览观光、文化体验、休闲娱乐等多功能的南中爨乡古镇。项目定位于"珠江源大城市西大门户",将其发展成为休闲度假胜地、现代文化新城。以三国风云和爨文化为故事主线打造一个历史文化旅游项目,并以特色旅游项目带动新城发展,打造集休闲、度假、居住、商务于一体的精品旅游文化示范工程和云南省新一轮旅游产业发展的新典范。

三大片区协同发力

规划将形成三大片区,其中现代新城片区将建立功能复合、多元的现代公共生活和服务中心;南中爨乡古镇片区(包含核心景区)将打造富有三国、爨文化特色的历史文化旅游标杆;休闲度假片区将做足温泉度假文章,汇集休闲体验、康体养生项目,打造西南地区最大的温泉度假、康体休闲度假片区。

The plan is to form Southern Central Cuan Xiang ancient town of multiple functions such as sightseeing, cultural experiencing, leisure and entertainment, with the local cultural features by focusing on the stories of the Three Kingdoms and the Cuan culture. The orientation of the project is "the west big gate of the big city of Zhujiangyuan River", to develop it into a recreational resort and a modern cultural town. Making use of the stories of the Three Kingdoms and the Cuan culture as the storyline to create a historical and cultural tourist program, and by using the tourism projects to propel the development of the new city, it will build a new model of high-quality tourism culture demonstration projects integrating leisure, vacation, residence and business and a new round of tourism industry development in Yunnan Province.

THREE LARGE AREAS IN COLLABORATION

The plan is to form three big areas, in which the modern new area will be built into modern public life and service center with comprehensive and diverse functions; the Southern Central Cuan Xiang ancient town area (including the core scenic spot) will be built into a benchmarking tourist attraction, rich in history and culture with the features of the Three Kingdoms and the Cuan culture; leisure resort area will make full use of hot-spring resort to create the largest hot-spring resort and recreational holiday area in the southwest region.

现状地形

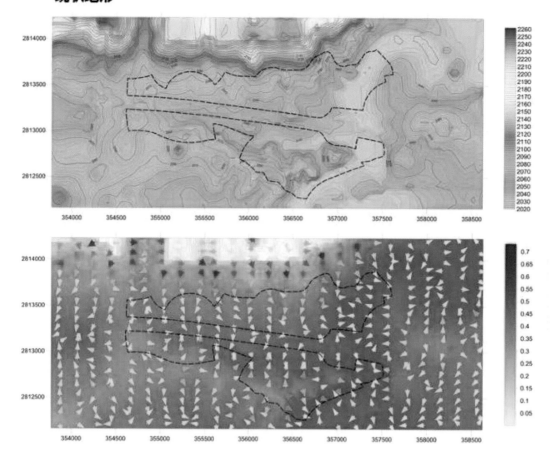

现状用地

现状旅游资源

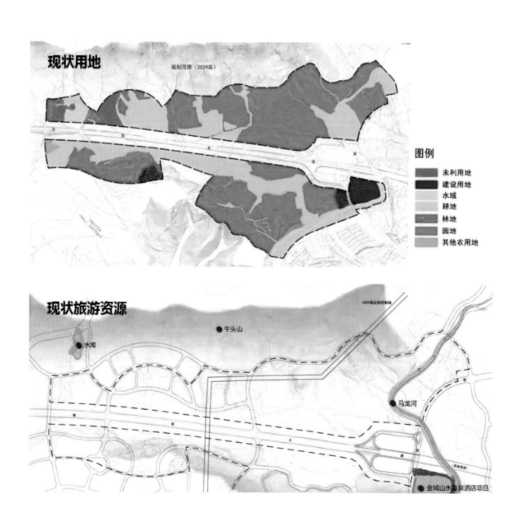

近期2013年 — 2015年
远期2016年 — 2020年

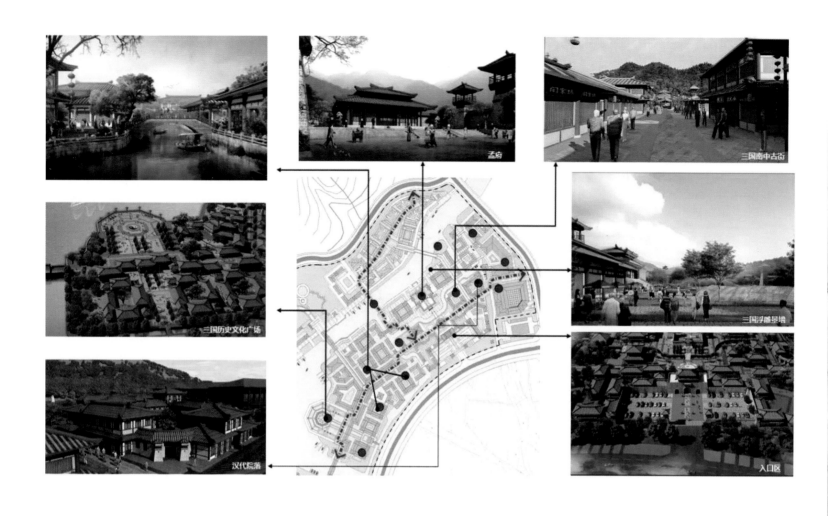
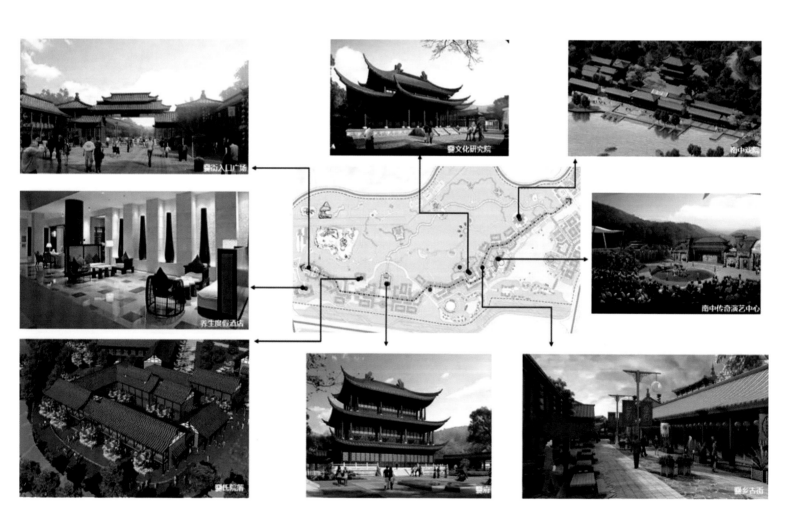

THE PLANNING AND DESIGN OF THE INTERNATIONAL TOURISM ISLET IN TANGSHAN BAY

唐山湾国际旅游岛规划设计

Location

河北省唐山市
Tangshan City, Hebei Province

Area

9.5 平方千米
9.5 square kilometers

Time

2012 年至今
Since 2012

本案的设计目标在于迎合滨海体验需求、对接区域一体、集聚活力产业、服务三岛旅游、突出特色魅力，打造最具吸引力的滨海城市核心区域。

策略详解

1. 以差异联动思路融入区域格局

以旅游发展、生态宜居为切入点，发展河北珍贵滨海空间，融入环渤海经济圈。

2. 以融入、驻留为主题的休闲旅游新思路

基地作为国际旅游岛的组成部分，应将旅游作为触媒，考虑具有国际化特征的高端居住功能和配套功能，以"留住人"为目的，形成"既是旅游区、更是健康生活区"的核心概念。

3. 滚动开发、周期调节的规划机制

基地发展目标宏大，在"开发一片，开放一片，适时调整，逐步实现"的原则下，结合国内外建设案例，探寻适合唐山湾的近、远期科学开发模式。

"一轴两心五片区"

本次规划为"一轴两心五片区"的规划结构，以乐北路为城市发展主要轴线，结合基地优质的滨海资源，打造蓝绿交错的"北方夏威夷"。

"一轴"：乐北路城市发展轴。乐北路是规划区联系南北交通的主要道路，由北至南串联国际旅游中心和商业商务中心，是城市形象展示的重要界面。

"两心"：国际旅游中心和商业商务中心。国际旅游中心位于基地最南端，尽享半岛三面环海之势，服务三岛旅游，构成旅游服务核心；商业商务中心位于规划区北部，提供区域内最前沿的游乐、购物、商务体验，构成新城核心。

"五片区"：规划区内不同的功能片区，包括创智产业办公片区、三贝码头综合服务片区、滨海养生养老居住片区、滨河高档居住片区、生态人文居住片区。

This project aims to create an attractive littoral city core area by catering to the demands of coastal experience, docking the regional integration, gathering dynamic industry, serving three islets tourism, and highlighting the charm of the local feature.

DESCRIPTION OF THE STRATEGIES

1. Regional framework integrated with differences' interaction

Starting with tourism development and eco-living, the strategy is to develop the precious coastal space of Hebei Province and to integrate it into Bohai-Rim Economic Circle.

2. Taking the new idea of leisure tourism themed on integrating and maintaining the residence

As a component of the international tourism islet, the base should take tourism as the accelerant, taking into account high-end residing and supporting functions with international characteristics, to work to "retain the residents", and to develop the core concept, "building it into a tourism and healthy living area".

3. Planning mechanism of rolling development and periodic adjustment.

This base has an ambitious development goal. Under the principle of "developing one area and opening up the area, with timely adjustment and gradual realization", it should combine construction cases both at home and abroad to explore short and long-term scientific development modes suitable for Tangshan Bay.

"ONE AXIS, TWO CENTERS AND FIVE AREAS"

The planned structure of this design is "one axis, two centers and five areas" and will use the Lebei Road as the main city development axis, to combine high-quality coastal resources of the base in order to create a blue and green "Northern Hawaii".

"One axis" refers to city development axis of Lebei Road. Lebei Road is designed to connect North-South traffic, connecting international tourism center and business center from the north to the south. It is an important interface to display the image of the city.

"Two centers" refer to international tourist center and business center. The international tourist center is located in the southernmost of the base, enjoys the sea on three sides of the peninsula, provides service for three islands and constitutes a core of tourism service center. The business center is situated in the planned area of the north, and it is to form a new town core which provides cutting-edge entertainment, shopping and business experience in this region.

"Five areas" refer to different functional planning areas including innovation industry area, Sanbei Port comprehensive service area, coastal health preservation and the elderly residential area, riverside luxury residential area and ecological and humanistic residential area.

① 未来广场
② 银厦北港蓝湾
③ 远大碧海蓝天
④ 水上飞行俱乐部
⑤ 环境监测站
⑥ 游艇俱乐部
⑦ 滨海商业街
⑧ 海之韵娱乐中心
⑨ 滨海大剧院
⑩ 温泉主题酒店
⑪ 西班牙风情商业街
⑫ 国际会议中心
⑬ 七星帆船酒店
⑭ 观海公寓
⑮ 海上嘉年华
⑯ 鹭鸶园
⑰ 会所
⑱ 海港集团置换用地
⑲ 水游城购物中心
⑳ 乐龄游乐中心
㉑ 商务产业中心
㉒ 财富中心
㉓ 医院
㉔ 五星级商务酒店
㉕ 酒店式公寓
㉖ 创智天地
㉗ 美食餐街
㉘ 海景嘉苑职工住宅小
㉙ 健康硅谷
㉚ 停车场
㉛ 精品购物中心
㉜ 生态酒庄
㉝ 滨海温泉主题社区
㉞ 健康养生主题社区
㉟ 居民安置区
㊱ 滨河生态社区
㊲ 生态公园
㊳ 三贝明珠码头
㊴ 索道站
㊵ 中学
㊶ 小学

THE DESIGN OF XINGDIAN CITY GROUP, THE TOWNSHIP OF WAGANG, QUESHAN COUNTY

确山县瓦岗镇镇区邢店组团城市设计

Location
河南省驻马店市
Zhumadian City, Henan Province

Area
9.9 平方千米
9.9 square kilometers

Time
2013 年至今
Since 2013

新型农村社区，指在农村区域按新型农村社区布局规划所建设的、居住方式与农村产业发展相协调，且具备完善基础设施和社会化公共服务设施的现代化新型农民聚居区。

新型农村社区空间发展规划，指在新型农村社区规划范围内，对产业发展、产业布局以及土地利用空间、基础设施和公共服务设施做出安排的规划。

打造一个亲水绿谷天堂，一处心灵康养小镇

以农村社区整合建设为契机，以产业发展为动力，以山水田园风光为特色，打造一个环境自然生态，生活设施完善，产业突出发展的宜工、宜居的新型社区形式，成为新型农村社区建设的样板。

规划深入挖掘规划区景观要素、结合产业发展趋势、发展创新要素，形成以下规划定位：

具有田园风光的生态宜居社区。依托现有环境、交通、人文资源，打造具有环境竞争力的田园宜居城区。

以康体养老为文化特征的养老社区。全力建设符合中国国情、贴合于老人生活模式的养老社区。

以发展商贸仓储物流为主导的重要产业基地。依托现有交通条件，形成先导产业清晰、核心产业明显、辅助产业完善的产业体系，使其成为地区商贸仓储物流中心，在更高的层面上获得经济发展动力。

以新型农村旅游为特色的体验区。借助基地周边旅游资源，引导地区经济模式转型，让它成为旅游服务产业和其他第三产业发展的重点基地。

A new rural community refers to the new modern farmers community constructed according to the new rural community layout and living pattern coordinating with the development of rural industries, and new and modernized residences for farmers equipped with complete infrastructures and socialized public service facilities.

The new rural community spatial development plan refers to the planning of industrial development, industrial layout and land use, infrastructures and public service facilities within newly planned boundary of the rural community.

BUILDING A WATERSIDE GREEN PARADISE, A TOWN FOR MENTAL AND PHYSICAL HEALTH

Seizing the opportunity of integrating and developing rural communities, driven by industrial development and with pastoral scenery as its feature, it will make a model for new rural community construction by creating a new type community with natural environment and ecology, complete living facilities, a highly developed industry with a comfortable place to live and work in.

The design is to dig deep into the landscape elements of the planned area, combine the industrial development trends and develop innovative elements, to form the following orientations:

This is to be an ecological and livability community that has idyllic scenery. Relying on the existing resource of environment, transportation and human resources, it will build an environmentally competitive pastoral livability city.

This is to be an elderly-care community featuring the culture of health and regimen. It will be committed to building a community caring for the elderly corresponding to China's actual conditions and fitting the living pattern of the elderly.

This is to develop the key industrial base that puts commercial warehousing logistics in a leading position. Based on the existing transportation condition, the industrial base is to establish an industrial system with distinct direction, explicit core industry and complete and sound supporting industry, to become the regional commercial warehousing logistics center and to get a driving force to a higher level for economic development.

This is going to be an experiencing area that features new type of rural tourism. With the support of tourism resources surrounding the base, it is to conduct the transformation of the regional economy model, making it a key base of tourism service industry and other branches of the third industry.

邢店组团

打造综合健康宜居、旅游休闲、康体养老、商贸仓储物流多元特色，成为区域可持续发展新型农村社区

Livable
健康宜居

完善的基础设施、建设高层次健康发展的

新型农村宜居小镇

Pension
康体养老

倡导绿色健康生活方式，建设

田园养老之地

Travel
旅游服务

依据地方特点，选择发展旅游服务、特色农业旅游服务等第三产业，形成

新型农村旅游热点

Logistics
商贸仓储物流

大力发展物流业，打造

商贸仓储物流中心

立足区域便捷交通条件，依托老乐山景区的生态自然资源

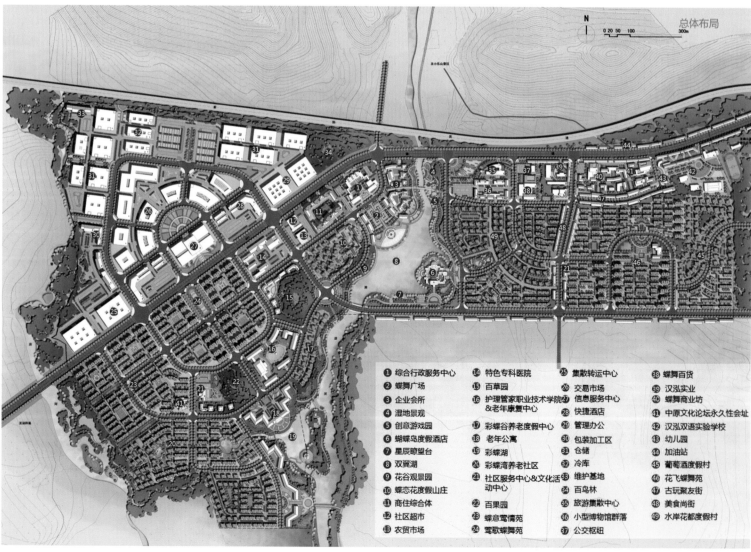

总体布局

① 综合行政服务中心
② 蝶舞广场
③ 企业会所
④ 湿地景观
⑤ 创意游戏园
⑥ 蝴蝶岛度假酒店
⑦ 星辰瞭望台
⑧ 双鬓湖
⑨ 花谷观景园
⑩ 蝶恋花度假山庄
⑪ 商住综合体
⑫ 社区超市
⑬ 农贸市场
⑭ 特色专科医院
⑮ 百草园
⑯ 护理管家职业技术学院&老年康复中心
⑰ 彩蝶谷养老度假中心
⑱ 老年公寓
⑲ 彩蝶湖
⑳ 彩蝶湾养老社区
㉑ 社区服务中心&文化活动中心
㉒ 百果园
㉓ 蝶意莺情苑
㉔ 莺歌蝶舞苑
㉕ 集散转运中心
㉖ 交易市场
㉗ 信息服务中心
㉘ 快捷酒店
㉙ 管理办公
㉚ 包装加工区
㉛ 仓储
㉜ 冷库
㉝ 维护基地
㉞ 百鸟林
㉟ 旅游集散中心
㊱ 小型博物馆群落
㊲ 公交枢纽
㊳ 蝶舞百货
㊴ 汉泓实业
㊵ 蝶舞商业坊
㊶ 中原文化论坛永久性会址
㊷ 汉泓双语实验学校
㊸ 幼儿园
㊹ 加油站
㊺ 葡萄酒度假村
㊻ 花飞蝶舞苑
㊼ 古玩聚友街
㊽ 美食尚街
㊾ 水岸花都度假村

THE RENEWAL DESIGN OF QUJING DOWNTOWN
曲靖市中心城区更新设计

Location	**Area**	**Time**
云南省曲靖市 Qujing City, Yunnan Province	0.57 平方千米 0.57 square kilometers	2013 年至今 Since 2013

设计团队整合了曲靖的发展优势，融入创新要素，结合对整体的研判，确定了本次规划的方案。

总体定位：区域之心、商业之都、繁华之城

经过改造，这里将成为曲靖具有划时代意义的城市中心区，涵盖主题性商业、五星级酒店、5A级写字楼、行政中心、高档公寓等多功能的复合型核心商业商务区。

"一个绿心四大板块"

本次规划确定了"一个绿心四大板块"的总体结构形式。"绿心"，即中心绿地，它是城市的绿色核心，体现麒麟山历史文化，是城市文化风貌的展示窗口；"四大板块"，结合"绿心"，集中布置城市综合体板块、行政办公综合板块、商业商务综合板块和公共服务设施板块。

The design team has integrated the development advantages of Qujing and innovative elements to determine the program of this plan through overall analysis.

GENERAL ORIENTATION: THE CENTER OF THE REGION, A BUSINESS CITY, A PROSPEROUS CITY

The redesigned area will become the city center of Qujing with epoch-making significance, a comprehensive core business district which covers theme businesses, five-star hotels, 5A-class office buildings, administrative centers, high-grade apartments and other functions.

"A GREEN CENTER WITH FOUR SECTIONS"

This plan determines the overall layout of "a green center with four sections". The "green center" means the green space in the center of the city, which is the green core of the city, embodying the history and culture of Qilin Mountain, and functions as the display window of the urban culture; the "four sections" together with the "green center" serve as a place where the four sections will be built: namely, the city complex, administrative complex, commerce and comprehensive commercial service and public service facilities.

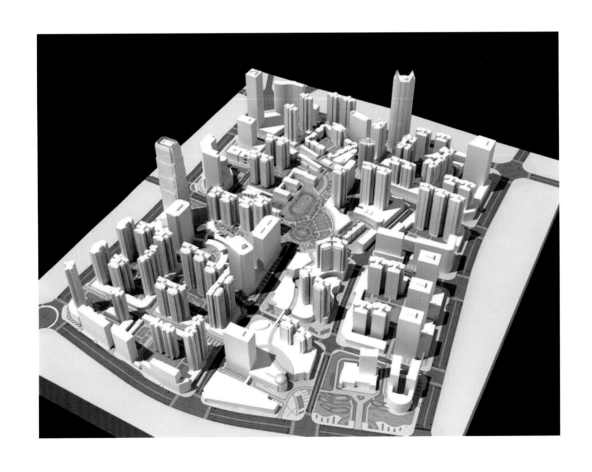

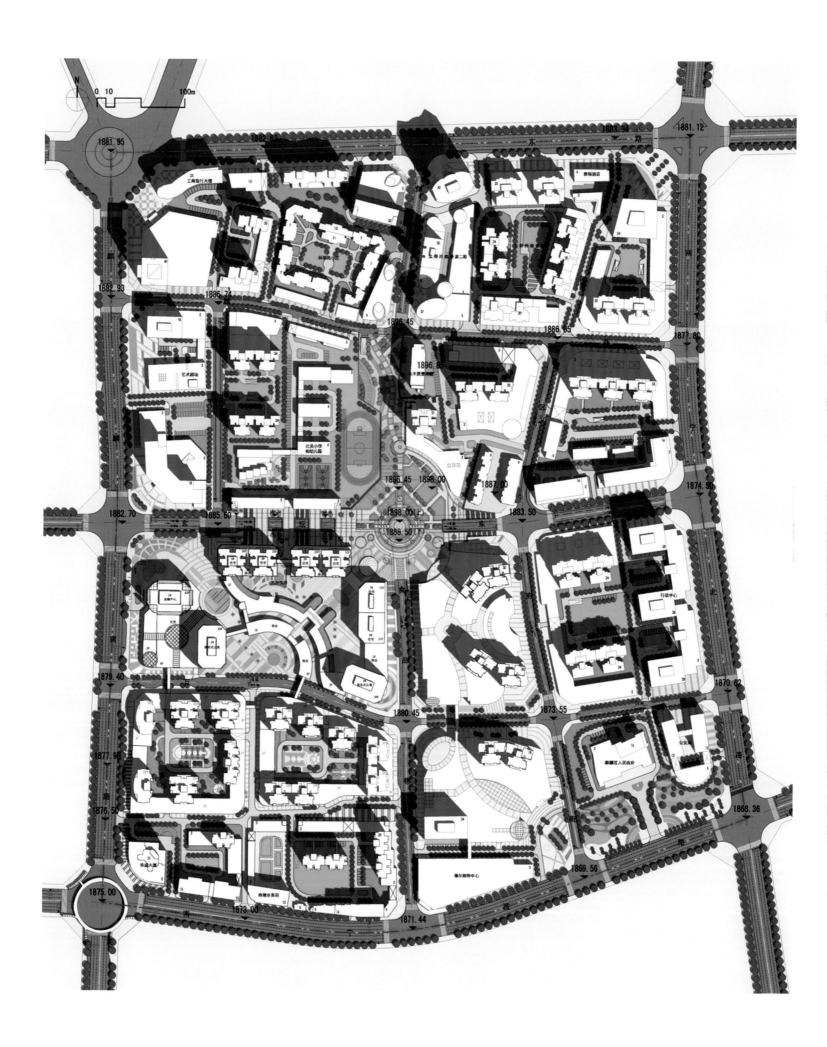

THE PLANNING AND DESIGN OF THE NEW TOWN OF DONGHU

东湖新城规划设计

Location

河北省邯郸市
Handan City, Hebei Province

Area
约 3.33 平方千米
About 3.33 square kilometers

Time
2013 年至今
Since 2013

项目依托文化、城市及生态资源，浓缩古赵文化精华，以建成极具吸引力的幸福宜居综合社区。

生态为景、生活为情、文化为体、现代为用

方案注重如下原则：

整体性原则：注重与周边区域、周边建设项目的协调互动关系，强调通过先进的开发理念，促进城市能级的提升。

生态性原则：城市生态环境的塑造，既是城市可持续发展的必然需要，城市高品质空间的要求，也是人们情感需求的基本元素之一。

多样性原则：居住、商业、休闲、旅游等各类功能的高度集约联动，是现代城市生活方式的反映，也是城市整体运营的高附加值所在。

标志性原则：优质的城市空间是城市营销的重要品牌载体，可以提高城市的辨识度，增强城市居民的认同感和归属感。通过具有地域特色和国际标准的标志性城市空间的打造，提升东湖新城地区乃至邯郸的城市核心竞争力。

The program is to create an attractive comprehensive community of happy livability, on the basis of cultural, urban and ecological resources, with the essence of the culture of ancient Zhao.

ECOLOGICAL LANDSCAPE, VIGOROUS ATMOSPHERE, CULTURAL EXPERIENCE AND MODERN FACILITIES

This project focuses on the following principles:

Principle of integrality: it emphasizes cooperation and interaction with the surrounding regions and construction projects around, as well as promotion of the urban function level through advanced developing concept.

Principle of ecology: the creation of urban ecological environment is not only necessary for urban sustainable development and high-quality life urban space but also essential for people's emotional need.

Principle of diversity: the highly intensive integration of functions in the aspects of residence, commerce, recreation, tourism, etc., reflects the modern urban lifestyle and high additional value of the overall city operation.

Principle of landmark: high-quality urban space is an important brand carrier for marketing, conducive to the promotion of urban identification and improvement of urban residents' sense of mutual acceptance and belonging. The creation of iconic urban space with regional characteristics and international standards can improve the core competence of the new town of Donghu and even of the city of Handan.

"十二连城，珠联璧合"

在基地中心规划一块圆形"璧玉"广场作为新城核心，在"璧玉"四周设置十二个组团，用绿带进行串联，形成"十二连城"。

- **一块 环形璧玉** 般的文化核心
- **十二座 价值连城** 的居住组团
- **一体 珠联璧合** 的公共开放空间

共同汇聚成财富集聚、文化地标的盛世赵都

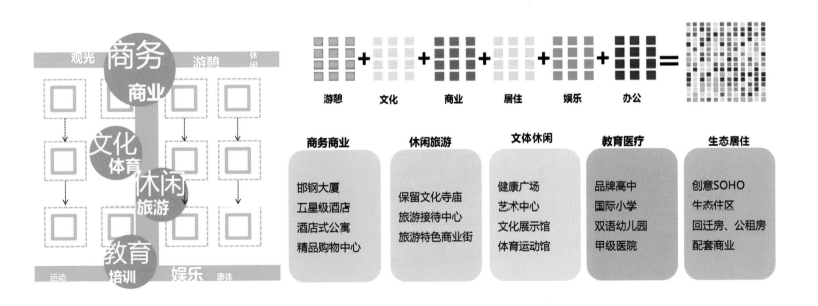

205

THE PLANNING AND DESIGN OF THE CREATIVE INDUSTRY TOWN IN DALI CITY

大理创意产业新城规划设计

Location

云南省大理市
Dali City, Yunnan Province

Area

约 2.04 平方千米
About 2.04 square kilometers

Time

2013 年至今
Since 2013

项目旨在打造中国滇西新派先锋产业集聚的生态文化智慧港。

三大策略全面推进

实现功能集聚，形成产业高地，延展、提升基地内部功能的基本门类，展现其商业、商务、旅游、展览、工业独特面，形成复合型创意产业集聚区。

营造特色街区，汇聚滇西和现代多种建筑风格与空间特色，打造极具地域标志性的城市空间。

彰显地域风貌，通过云南、泰国等地域文化、休闲文化等多元文化的注入，实现项目的多元提升。

The project aims to forge an ecological port with culture and wisdom assembled with new Chinese Yunnan Pioneer Industries.

THREE STRATEGIES ARE COMPREHENSIVELY PROMOTED.

To achieve functional agglomeration, form industrial highlands, extend and improve the basic classes of the basic internal functions, this is to form a comprehensively creative industry cluster demonstrating its unique features in business, commerce, tourism, exhibition, industry.

It will create a special block, by gathering a variety of architectural style and space characteristics of Western Yunnan and modern style, to forge an urban space of regional landmark.

To highlight regional style, it is to upgrade multiple factors of the program by integrating the regional and leisure culture of Yunnan and Thailand, etc.

山——维护自然和谐循环

菠萝山的地形和地貌是基地最重要的生态脉络和规划结构。

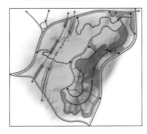

绿——网罗葱茏的城市生活

依山体和山谷形态的绿网连缝的开放自然生态空间系统，将山体景观延续到城里。

人——带来交流和融合

道路和公共交通把四面八方的人带到智慧之城。

活力——孕育新的文化和生机

建立不同的功能，开放空间和公共交通相串联，创造不断的交流与互动。

THE PLANNING AND DESIGN OF DONGJIN INTERNATIONAL TOWN

东津国际小镇规划设计

Location	Area	Time
湖北省襄阳市 Xiangyang City, Hubei Province	1.11 平方千米 1.11 square kilometers	2013 年至今 Since 2013

东津国际小镇致力于打造云智慧社区生活平台，开创全球城居新模式，成为引领智慧生活新时代的智慧城邦，以及融汇国际前沿品质生活的国际社区。

碧水湾、黄金道、精英城，全新城居新模式

这不仅仅是一个生活之城，还是融合了云计算、互联网、通信网和物联网的云居住社区，它超越了传统社区概念，将开创出一种城居新模式。

基于对新城市主义居住的深刻理解，规划方案由不同院落街区组成生活城，全面实现新城市主义的生活理想，并通过提供居住、生态、健康、商业、服务、交通等六种A级居住体验为居住者的生活带来惊喜。

Dongjin International Town is committed to building a cloud wisdom community life platform, creating a new model of global urban life, becoming a wisdom city leading a new era of wisdom life, and an international community blending international forefront quality life.

CLEAR WATER BAY, GOLDEN ROAD, ELITE CITY, NEW MODE OF A BRAND-NEW CITY LIFE

It isn't merely a city of ordinary life, but a cloud community that integrates cloud computing, the Internet, communication network and Internet of things. It surpasses the traditional concept of the community, and will usher in a new model of urban living style.

Based on a deep understanding of the new urbanism living, the planning arranges different courtyards and blocks into towns to realize the ideal living of new urbanism, with happy surprises brought to the residents by the six A-level living experiences such as housing, ecology, health, business, service, and transportation.

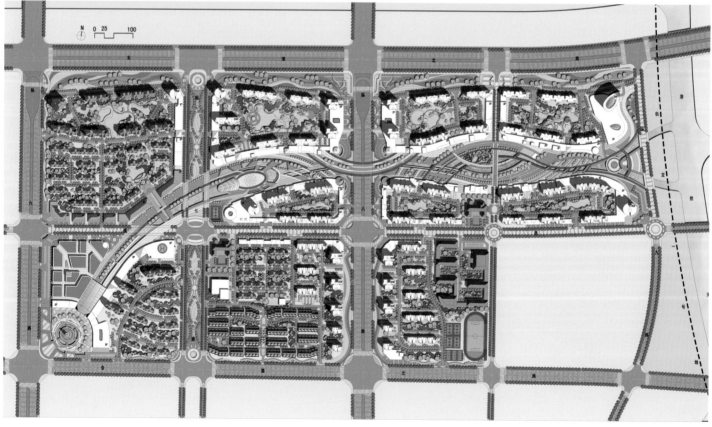

THE PLANNING AND DESIGN OF THE SOUTH OF UNIVERSITY ROAD

学府路南侧新城规划设计

Location	Area	Time
安徽省蚌埠市 Bengbu City, Anhui Province	0.45 平方千米 0.45 square kilometers	2013 年至今 Since 2013

项目旨在打造一个"城市磁极、活力容器、核心地标"。

充满活力的商务休闲生活空间

营造"绿色生态"的办公空间，以开放、通透的格局引入明亮充足的阳光。规划设计精简现代的细部，强调形式服务于功能，形成一个纯粹的商务休闲生活空间。另外，还配备了大尺度空中花园。建筑布置有序，高低错落，内外相连，收放自如，围合但不失开敞。真正全方位地引入项目周围的自然景观，内景和外景合二为一。

The aim of the project is to forge a "city magnetic pole, dynamic center and core landmark".

COMMERCIAL AND LEISURE LIVING SPACE OF VITALITY

This is to create an office space with "green ecology", to introduce bright and adequate sunshine with an open, transparent pattern. The planning is to streamline the details of modernity, emphasizing that the form serves the function, so as to form a purely commercial and leisure living space. In addition, it is to be equipped with a large-scale aerial garden. The building will be arranged in an orderly manner, with the high and the low, zigzagging, connecting the inside and outside, encircled but open enough. It truly introduces natural landscape nearby all-rounded, achieving the effect of integration the interior and exterior scenery.

三大策略，相辅相成

设计提出三大策略：

1. 开启绿色生态、人性化的花园办公时代。

2. 打造城市印象新标志，引领蚌埠都市化进程。

3. 多功能综合开发，建立物业生态平衡系统。

THREE STRATEGIES COMPLEMENTING EACH OTHER

Three strategies are proposed:
1. To initiate a green, ecological, and humanity garden office era.

2. To create a new logo of the city impression and lead the process of urbanization of Bengbu.

3. To establish ecological balance system with multi-functional and integrated development.

THE PLANNING AND DESIGN OF YUDU GONGJIANG RIVER WATERFRONT NEW CITY

于都贡江滨水新城规划设计

Location	Area	Time
江西省赣州市 Ganzhou City, Jiangxi Province	0.33 平方千米 0.33 square kilometers	2013 年至今 Since 2013

在美丽的江西赣州，正待崛起的于都贡江滨水新城将被打造为地标性的商务中心、24小时娱乐中心，商业商务双核并举的特色商业街区。

规划理念为滨水生态、文化记忆、低碳高效、多元复合。

功能定位

贡江新城的功能定位为于都"乐居、乐业、乐游"的先行区，打造"城市综合中心、生态宜居、休闲游憩、地域文化弘扬"的城市职能。城市主题确立为体验（市民与场所互动）、文化（城市与品牌互动）、多元（空间与功能互动）。

主题形象

通过对贡江新城形象需求的塑造、功能需求的提升、产业需求的升级、空间需求的拓展、旅游需求的提升、房地产需求的促进这六大举措，展示"生态之城、活力之城、美丽之城"的城市形象。

功能结构：一心两轴两片三带

根据"两条轴线、三级结构、四大支撑、五种风情、六核驱动"的规划构思，形成如下的结构。

"一心"：以水南大道与长征大桥南延路交会区域为整个规划区的中心，它同时也是于都县中心城区的主中心。

"两轴"：依托水南大道和长征大桥南延路，形成两条城市空间拓展的主轴线。

"两片"：规划区以长征大桥南延路为界形成两大功能片区。南延路以西为行政商务综合区，布置未来于都县的行政中心及经济区、金融商务区等；南延路以东为商业娱乐综合区，布置大型的城市综合体、商业慢行系统、文化娱乐设施、医疗卫生设施等。

"三带"：围绕贡江南岸、结合自然山体连通贡江与楂林河，保留并强化于都县自老县政府至罗田岩的城市轴线，构筑三条贯穿贡江新城区域的绿化带，打造城市绿道系统，构建良好的城市自然生态环境，为市民提供良好的生活、游憩场所。

In the beautiful Ganzhou, Jiangxi, Yudu Gongjiang River waterfront new city is rising, and it will be built into a landmark of business center, 24 hours entertainment center, and dual cores of commerce and commercial service.

Planning ideas are waterfront ecology, cultural memory, low carbon efficiency, multiple integration.

FUNCTION ORIENTATION

The city is determined to have the function of the Yudu leading district of "happy living, happy career, and happy touring", and satisfy the urban function of "city complex center, ecological livability, leisure and recreation, and dissemination of regional culture". The theme for the city is experience (interaction of the residents and the places), culture (interaction of the city and the brands) and multiplicity (spatial and functional interaction).

THEMATIC IMAGE

Through six measures of shaping the Gongjiang River waterfront new city image, improving functions, upgrading the industry, expanding the space, satisfying the demand of the tourism and promoting the demand of real estate, the project will display the image of "city of ecology, city of vitality, city of beauty".

FUNCTIONAL STRUCTURE: ONE CENTER +TWO AXES +TWO DISTRICTS +THREE BANDS

According to the planning concepts of "two axes, three levels of structures, four supports, five styles and six drives", the following structures are formed:

One center: the center of this planned area is going to be the intersection region between Shuinan Avenue and the southern extending road of Changzheng Bridge, which is also the main center of the down town of Yudu.

Two axes: two main axes for urban space extension are formed based on Shuinan Avenue and the southern extending road of Changzheng Bridge.

Two districts: the planned area is to form two main functional districts separated by the southern extending road of Changzheng Bridge. To the west of the southern extending road is going to be administrative-commercial comprehensive district, to function as future administrative center, economical area, financial and business areas of Yudu. To the east of the southern extending road is going to be a commercial-entertainment complex, to be developed for large urban compound, slow commercial traffic system, cultural entertainment infrastructure and medical infrastructure,etc.

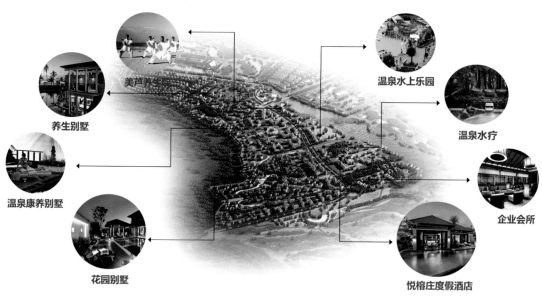

Three bands: around the south bank of the Gongjiang River, combined with the natural mountain and connecting Gongjiang River and Zhalinhe River, retaining and strengthening the city axis of the Yudu County, from the old county government to Luotianyan, it is to build three green bands passing the new city area of Gongjiang River waterfront new city, in order to build the city greenway system, to give the city a beautiful natural environment, and to provide the citizens with ideal place for living and recreation.

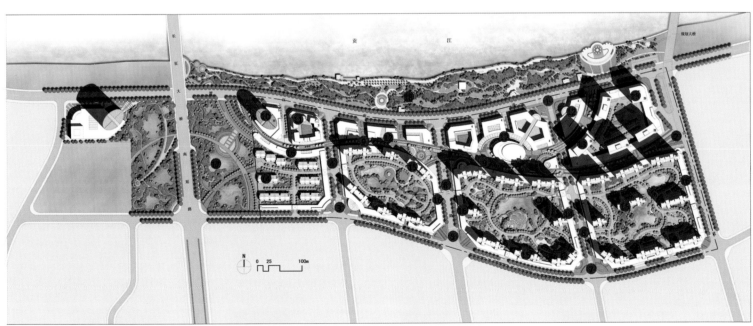

- ① 商业街入口广场
- ② 商业街东侧主入口
- ③ 休闲娱乐街入口广场
- ④ 休闲娱乐街
- ⑤ 饕餮饮食街入口广场
- ⑥ 饕餮饮食街
- ⑦ 时尚艺术街入口广场
- ⑧ 时尚艺术街
- ⑨ 步行商业街
- ⑩ 酒店办公
- ⑪ 办公
- ⑫ 商业
- ⑬ 公寓
- ⑭ 花园洋房
- ⑮ 高层住宅
- ⑯ 底商
- ⑰ 濂溪公园
- ⑱ 贡江南岸路堤景观公园

THE PLANNING AND DESIGN OF NEW LAISHUI COUNTY

涞水县新城规划设计

Location	Area	Time
河北省保定市	4.54 平方千米	2014 年至今
Baoding City, Hebei Province	4.54 square kilometers	Since 2014

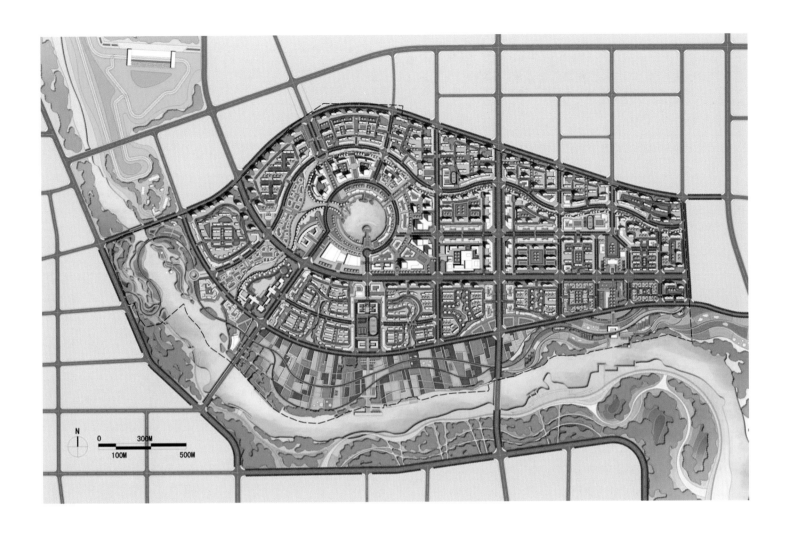

"望得见山、看得见水"，是人们对于城市环境共同的渴望。涞水新城，有着真山、真水。

"一带两轴三区"的布局

"一带"：滨水风情景观带。

"两轴"：商业景观轴、艺术景观轴。

"三区"：北翼游乐休闲区、核心京涞新都心、南翼创意先锋区。

THE LAYOUT OF ONE BELT, TWO AXES AND THREE AREAS

"With mountain and river in sight" is people's common desire for the urban environment. New Laishui county is blessed with authentic mountains and rivers.

"One belt" refers to the waterfront landscape belt.

"Two axes" refer to business landscape axis and artistic landscape axis.

"Three areas" refer to the northern recreation area, new city heart of Jinglai core, and southern creative pioneer area.

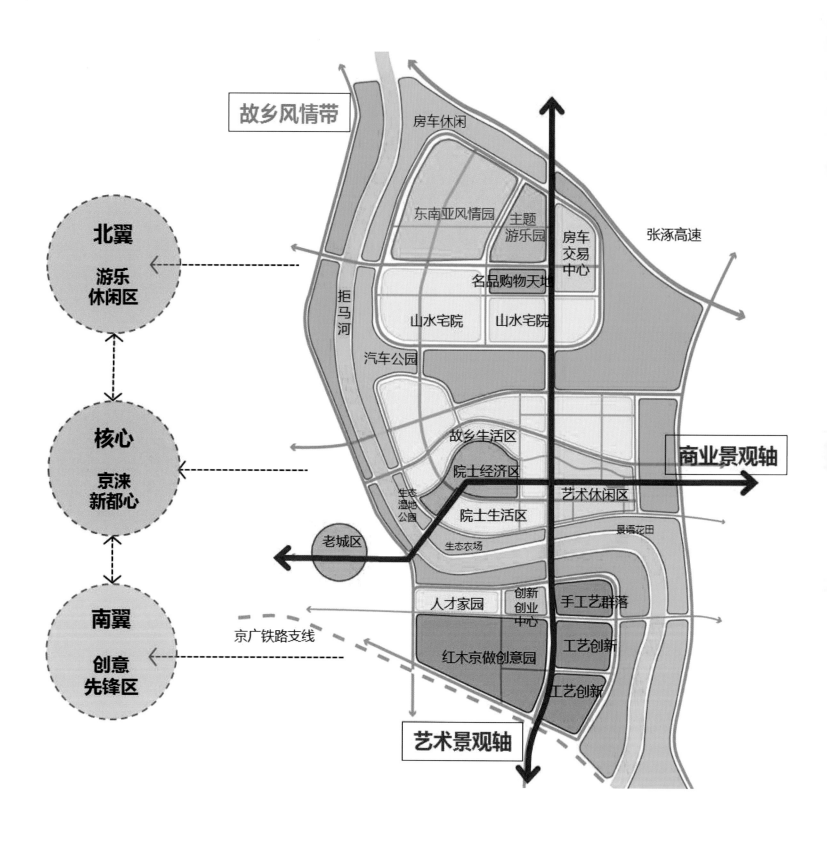

THE PLANNING AND DESIGN OF THE NEW INTERNATIONAL TOWN OF TOURISM VACATION IN BAIYANGDIAN

白洋淀国际旅游度假新城规划设计

Location

河北省保定市
Baoding City, Hebei Province

Area

约 14.1 平方千米
About 14.1 square kilometers

Time

2014—2025 年
2014—2025

本次规划综合分析了白洋淀旅游发展的机遇和困境，在资源梳理和区域整合的基础上对接市场需求，确定了白洋淀是集生态休闲、文化民俗体验、温泉康疗养生、田园度假、湿地观光等为一体的世界级休闲度假旅游胜地。

宏观格局

基地的使命是立足资源特征，开发特色旅游体验项目，将白洋淀国际旅游度假区打造为世界级休闲度假旅游目的地，使之成为一个美丽中国的生态典范、一个聚水汇贤的国际高端会客厅、一个华北首席旅游目的地和集散地，成为京津冀城市群生态坐标、城乡统筹发展新标杆、京津冀旅游升级"新示范"、温泉度假生活"新极核"。规划的愿景是打造生态坐标、代言文化精神、构建旅游核心。

项目的基本市场面可分为三个能级，其中基础市场／一级目标市场为京津冀地区客源市场，主体市场／二级目标市场为邻近省份和地区，机会市场／三级目标市场为国内其他地区市场。

The planning analyzes comprehensively the opportunities and straits in the development of the tourism in Baiyangdian. The plan that is based on resource sorting and area integrating to dock market demands, has decided to build Baiyangdian into a world class leisure resort integrating ecological leisure, cultural folklore experience, SPA health care, pastoral holiday, wet land tourism, etc.

MACROSCOPIC CONFIGURATION

The mission of the base is to develop the featured experience program of tourism project according to the characteristics of resource to forge Baiyangdian international tourism resort to a world-class leisure destination, so as to become a beautiful ecological model of beautiful China, an international high-end reception hall, a chief tourist destination and distribution center in North China, to become the ecological coordinate of Beijing, Tianjin and Hebei city group, the new benchmark of the coordinated urban and rural development, the "new model" of Beijing, Tianjin, and Hebei tourism and "new polar core" of SPA resort. The vision of the plan is to forge the basic market coordinates, to represent culture spirit, and to construct the tourism core.

The basic market of the program can be divided into three levels, among which the basic market — first level market is the Beijing, Tianjin, and Hebei region customer market, the main market — second level market is the neighbouring provinces and regions, and the opportunity market — third level market is other domestic market areas.

总体定位

规划形成"一核两翼三区"的旅游发展格局,重点开发"一核"——人文水镇,延伸两翼的"绿色核心"湿地公园与"文化核心"渔农之乡,打造特色旅游产品——"温泉康养区"、"运动休闲区"和"白洋淀区",形成多功能联动的复合型旅游度假区。

产品定位

规划确立三大核心产品,即人文水镇、渔农之乡及湿地公园。主题产品包括水韵商街、芦音剧院、非遗博物馆、温泉度假酒店、市民亲子农园和湿地荷花淀等。

总平面及重点产品布局

通过湿地修复示范、保护性利用,还原白洋淀肌理,重现芦荡十八湾风貌。形成原生芦苇荡、湿地科普、湿地净化、湿地探索、西淀荷风等五大片区。同时,还将再现"湿地八景"景观:美芦渔家、鹭洲鸥渚、西淀荷风、芦海水韵、秋水归雁、鹅湖映霞、观鸟天堂、渔家野渡。将生态湿地公园打造成为全国湿地治理示范区、湿地保护性开发样板区和华北湿地保护教育基地。

GENERAL ORIENTATION

This plan is to form a tourism development pattern of "one core, two wings and three districts", and attaches importance to "one core" development — water town with a sense of humanity, extends the two wings — wetland park as "green core" and the village of fishing and farming as "cultural core", and creates special tourist products, such as hot spring regimen therapeutic baths, sports and recreational area and Baiyangdian water area, forging an integrated tourist and holiday resort with multiple functions.

PRODUCT ORIENTATION

The plan establishes three core products, including water town with a sense of humanity, village of fishing and farming, and wetland park. Theme products consist of water-rhyme business streets, Luyin (reed sound) theater, intangible cultural heritage museum, hot spring resort hotel, and parents-kids farm garden, as well as lotus creek.

THE GENERAL ARRANGEMENT AND LAYOUT OF KEY PRODUCTS

The planning aims to restore the original scenic features of Baiyangdian and its landscape of zigzagged reed marshes by means of wetland restoration demonstration and protective measures. In this way, five main sectors are to be formed including the Original Reed Marshes, Wetland Science, Wetland Purification, Wetland Exploration and West Lotus Flower Lake accordingly. At the same time, the "Wetland Eight Scenes" will also be reproduced: the Fishermen in the Reed, Heron Islands, West Lotus Flower Lake, the Rhyme of the Reed Sea, Autumn Home-coming Wild Geese, Swan-lake with Evening Glow, Bird-watching Paradise, Fishermen's Wild Ferry. The ultimate goal is to build the Ecological Wetland Park as the national demonstration area of wetland restoration, the sample areas of the protective development of the wetland, and the North China education base of wetland protection.

01 小镇规划馆
02 渔人码头
03 运河水岸天地
04 芦音剧场
05 河畔温泉度假酒店
06 白洋淀婚俗馆
07 皇家行宫展览馆
08 静修书院
09 滨水主题客栈
10 爱的主题体验区
11 河鲜美食街
12 五彩花田
13 地热温泉农场
14 安纳伯格庄园
15 温泉度假村
16 综合娱乐馆
17 温泉休闲中心
18 温泉大观园
19 水生植物观赏园
20 高尔夫体育公园
21 悦榕庄度假酒店
22 花园别墅
23 温泉康养别墅
24 养生别墅
25 温泉养生会所
26 温泉水上乐园
27 美芦养生院
28 温泉SPA
29 美芦高尔夫球会
30 湿地公园泄洪区
31 湿地植物园
32 湿地八景
33 原生态芦苇区
34 荷花淀
35 码头集散中心
36 滑翔伞
37 热气球
38 湿地博物馆

● 主入口

THE PLANNING AND DESIGN OF SHANGHAI POLAR SEA WORLD

上海极地海洋世界规划设计

Location	Area	Time
上海市浦东新区 Pudong New Area, Shanghai	约 0.30 平方千米 About 0.30 square kilometers	2014 年至今 Since 2014

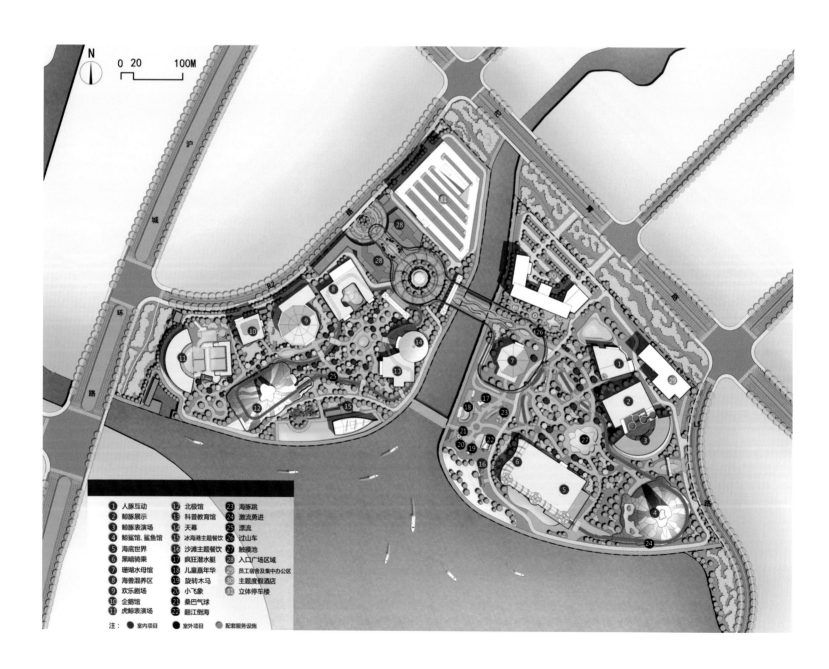

上海极地海洋世界所在的区域是上海浦东新区的临港新城主城区边缘，对外交通条件较好。

海洋世界，快乐体验

项目定位于打造以海洋文化为核心的区域度假品牌，用"生态关爱"来推广"快乐精神"，创建可持续生命力的旅游产品体系，建立形式与内容并重的"情景式体验度假胜地"。

Shanghai Polar Sea World is located at the edge of main urban area of Lingang New City, Pudong New Area, Shanghai, with convenient transportation conditions.

SEA WORLD, FUN EXPERIENCE

This project aims to create a regional resort brand centering on "sea culture". Using "eco-friendliness" to popularize "happiness", it will build sustainable tourism product system, and establish "situation-oriented experience resort" stressing both form and content.

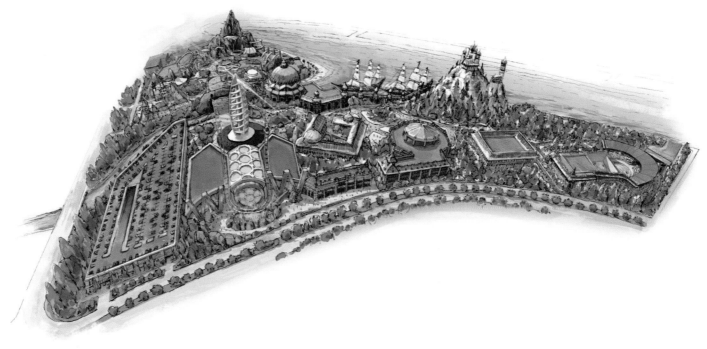

239

THE PLANNING AND DESIGN OF THE NEW HIGH-TECH AREA, ZHUOZHOU

涿州高新技术产业新城规划设计

Location	Area	Time
河北省涿州市 Zhuozhou City, Hebei Province	95.2 平方千米 95.2 square kilometers	2014 年至今 Since 2014

规划以涿州独有的"义"文化为核心，以自然的生态环境、高效的产业经济和多元的休闲旅游为支撑，打造京津冀协同发展示范区，国际化的创新型生态智慧未来城。

项目寻求产业的一体化发展、承接产业转移、实现错位发展、遴选引擎产业，同时，将构建四大提升产业、发展三大支撑产业。

"一环两核四心五片"

"一环"：绿色活力环，通过构建兼具交通及生态功能的城市绿色活力环，串联各个片区及区域核心。

"两核"：城市魅力核、生态绿核。通过构建区域的城市CBD，以充满活力的商业、商务、会展功能汇聚城市能量，发展成为未来本区域的魅力之核。结合内部固有资源，联系周边生态绿地，在基地内部打造生态绿核，注入生态绿色元素。

"四心"：文化中心、科教中心、康养旅游中心、商贸中心。它们是各具特色的次级功能核心，形成多中心模式，激发区域活力。

"五片"：城市综合片区、仓储物流片区、高新产业片区、国际养生片区和教育科研片区。

The planning is to center on the exclusive righteousness culture, supported by natural eco-environment, high-efficient industrial economy and multiple leisure tourism, and aims at building a co-developing demonstration area of Beijing-Tianjin-Hebei Region, and an international innovative eco-wisdom future city.

This project pursues integrated development of industries, to carry out industry transfer, to realize misplacement development and to select engine industries. Meanwhile, it will form four promoting industries and develop three supporting industries.

"ONE RING, TWO CORES, FOUR CENTERS AND FIVE AREAS"

"One ring" is green vigor ring. By building a city green vigor ring serving both traffic and ecological functions it will link up each area and regional core.

"Two cores" are a charming urban core and a green core. By forming a regional city CBD, and gathering city energy with active business, commerce and exhibition functions, this place will bloom into an enchanting core of this region in the future. Combining existing internal resources and linking periphery green lands, a green core will be built in the base, thus introducing the green element.

"Four centers" are cultural center, science and education center, health and caring, tourism center as well as commerce and trade center. They are secondary functional cores with respective characteristics, which form a polycentric mode and stimulate the regional vitality.

"Five areas" are city comprehensive area, storage and logistics area, high-tech industry area, international health-preserving area as well as educational and scientific research area.

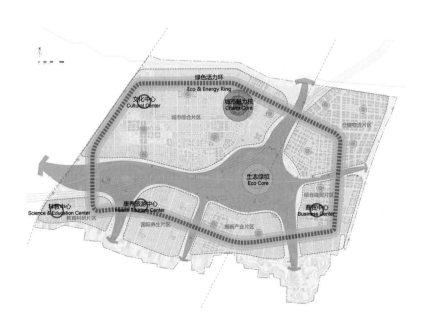